Global Climate Change Effect on Agriculture Production

Global Climate Change Effect on Agriculture Production

Nehal Kidwai

RANDOM PUBLICATIONS
NEW DELHI - 110 002 (INDIA)

Global Climate Change Effect on Agriculture Production

ISBN 978-93-51112-79-2

Published in 2014 in India by

RANDOM PUBLICATIONS

4376-A/4B, Gali Murari Lal, Ansari Road
New Delhi-110 002
Phone: +9111-43580356, 23289044
E-mail: randomexports@gmail.com; sales@randompublications.com; info@randompublications.com

Reprinted 2026

Type Setting by: Friends Media, Delhi-110089
Digitally Printed at: Replika Press Pvt. Ltd.

Preface

The risks associated with climate change lie in the interaction of several systems with many variables that must be collectively considered. Agriculture (including crop agriculture, animal husbandry, forestry and fisheries) can be defined as one of the systems, and climate the other. If these systems are treated independently, this would lead to an approach which is too fragmentary. The issue is more global. It is now held as likely that human activities can affect climate, one of the components of the environment. Climate in turn affects agriculture, the source of all food consumed by human beings and domestic animals. It must be further considered that not only climate may be changing, but that human societies and agriculture develop trends and constraints of their own which climate change impact studies must take into consideration. In general, global food production has been growing faster than human population. There are, however, marked disparities between continents. For example, in Africa local production of cereals cannot keep pace with population increase, and production of root and tuber crops is growing faster than that of the nutritionally more valuable cereals. In contrast, arable land growth lags behind population growth, which indicates some intensification of production. In Asia, the upper limit of available land has been reached in several countries, resulting in very high cropping intensities and a dominant role for irrigation. In Latin America, the increase in arable land is achieved only at a high ecological cost (especially deforestation) which may have direct relevance to climate change. Climate constitutes a complex of inter-related variables. On average, through a set of regulatory mechanisms, a smooth change in one variable triggers smooth changes in most others. With the exception of possible qualitative and abrupt variations, which will be mentioned below, such inter-relations are independent of atmospheric carbon dioxide (CO_2). The latter and other greenhouse gases play a part largely through their effect on the radiation balance of the atmosphere.

There is only a weak link between such factors as cloudiness and wind. Temperature, evaporation and rain are strongly correlated, which illustrates the likely intensification of the hydrological cycle. Combined with the projected pressure on land and water use, competition for land and water will certainly become a key social and political issue. Climate variability is likely to increase under global warming, both in absolute and in relative terms. This is linked with thresholds which affect the occurrence of many meteorological phenomena. For instance, tropical cyclones are 'fed' by water vapour evaporating from oceans at a temperature above 26 or 27°C. Therefore, higher average sea surface temperatures are bound to result in a higher frequency of tropical cyclones. The rate of change itself is extremely important. For example, recent work on the saw-tooth temperature changes in the past as observed in the Arctic, raise concern that changes may occur abruptly, with average temperatures changing by 10 or 12°C in just a matter of decades. The mechanism of such changes is not clear as yet, but seems to involve the mechanical stability of ice sheets, and sudden changes in the hydrological cycle and the Atlantic conveyor current.

The contents of the book will help scientists and academics to examine, modify and improve of this subject.

I thank all members of my team who have helped in the preparation of the book. My special thanks go to "Random Publications" who have published the book.

—Nehal Kidwai

Contents

1

Global Climate Change for Agriculture and Trade

Until recently, the climate change adaptation literature has tended to downplay the impacts on agriculture. Two different effects – CO_2 fertilization and changes in trade flows – were together assumed to offset much of the negative effects of climate change. Experiments in the laboratory show that higher CO_2 levels increase yields, especially of the so-called C3 plants such as rice, wheat, soybeans and sorghum.

The "positive effect of trade" logic relies on the assumptions that changes in trade flows will allow exploitation of changing comparative advantage brought about by climate change and that trade liberalization might further reduce the costs. Agricultural trade flows depend on the interaction between comparative advantage in agriculture, which is determined by climate and resource endowments, and a wide-ranging set of local, regional, national and international trade policies. Crop and animal production is affected by changes in temperature and precipitation. Because climate change results in new patterns of temperature and precipitation, agricultural comparative advantage also changes, setting up the possibility of changes in trade flows as producers respond to changing constraints and opportunities.

As with any change in comparative advantage, unfettered international trade allows comparative advantage to be more fully exploited. Restrictions on trade risk worsening the effects of climate change by reducing the ability of producers and consumers to adjust. It is also important to point out that if climate change reduces productivity of some crops in some regions and doesn't increase productivity adequately in other regions, trade cannot fully compensate for the global reduction in productivity. Climate change projections indicate that if temperature increases are severe enough, a global net reduction in productivity will be unavoidable.

Uncertainties in future climate outcomes make it difficult to determine the effects on agricultural productivity, and therefore world trade flows. The uncertainties of future agricultural policy regimes make simulations doubly uncertain. Nonetheless, some researchers have attempted to do so. Negative yield effects in temperate grain producing regions would be buffered by interregional adjustments in production and consumption and corresponding trade flows. A key assumption was that part of the production losses from temperature and rainfall would be offset by CO_2 fertilization. Another key result was that agricultural trade flows would support an agricultural system relatively resilient in the face of uncertain effects of climate.

A widely cited 2004 publication based on more complex modelling of both climate and agriculture using the climate modelling results of the IPCC's Third Assessment was still relatively sanguine about global food production but with more caveats than earlier chapter. "... the combined model and scenario experiments demonstrate that the world, for the most part, appears to be able to continue to feed itself under the SRES scenarios during the rest of this century. The explanation for this is that production in the developed countries generally benefits from climate change, compensating for declines projected for developing nations. While global production appears stable, regional differences in crop production are likely to grow stronger through time, leading to a significant polarisation of effects, with substantial increases in risk of hunger amongst the poorer nations, especially under scenarios of greater inequality".

These results are strongly influenced by the assumed CO_2 fertilization effect of over 10 percent for wheat, rice and soybeans and five percent for maize. Without CO2 fertilization, the prognosis is not nearly so bright. A 2007 study (J. Reilly et al., 2007) that simulates agricultural response to climate change and incorporates general equilibrium economic effects finds that yields would likely increase in all regions, with smaller gains in the temperate regions than previous models but positive yield changes in the tropics. As with the earlier studies, their results are strongly affected by the CO2 fertilization effect. In addition, they make fairly strong assumptions about crop biological behaviour in response to climate and other changes.

Two important questions stand out when evaluating these studies. First, the benefits of CO_2 fertilization are extremely important in essentially mitigating the rainfall and temperature effects of climate change. As mentioned above, the CO_2 fertilization effect works most strongly with C_3 crops. The two most important food crops – rice and wheat – use C_3 photosynthesis, as do soybeans and potatoes. Maize, sorghum, millets, and sugar cane are examples of important crops that use C_4 photosynthesis and where the fertilization effect is smaller, even in the laboratory.

Recent field experiments on CO_2 fertilization, find that the effects in the field are approximately 50 percent less than in experiments in enclosed containers. And another report finds that higher levels of atmospheric CO_2 increase the susceptibility of soybean plants to the Japanese beetle and maize to the western corn rootworm. So the actual benefits of CO_2 fertilization in farmer fields remain uncertain.

Second, the results in the earlier literature all depend on a relatively open world trading system where climate-induced shortfalls in some regions can be offset by imports from others. The recent lack of progress in the Doha Round and significant trade restrictions imposed during the 2008 food price crisis suggests that we should not be sanguine about the role of trade flows in agricultural adjustments to climate change. The recent disruptions in trade and in food availability with sharp price increases highlight the fragility of the food system in many poor countries, and its vulnerability to the kinds of variations in production that many predict with climate change. Finally, these studies tend to focus on staple food crops, but the recent history of agricultural trade is driven by the rapid growth in production and export of high-value agricultural crops from the developing world, often produced in niche agro-climatic zones. These exports have provided part of the foreign exchange needed to allow developing countries to import the food and feed demanded with growing incomes. Essentially no research has been done on the extent to which those products would be affected negatively or positively by climate change. One could imagine, however, that sea level rise would negatively affect developing-country exports of seafood (in particular, shrimp raised in low-lying ponds) and relatively small temperature increases would affect temperate crops such as horticulture crops grown in niche environments elsewhere. This is clearly an area where new research is badly needed. A recent study (Gerald C. Nelson et al., 2009) and related research conducted at the International Food Policy Research Institute (IFPRI) provides the most current evidence on the effects of climate change on agriculture.

Because climate change simulations are inherently uncertain, two climate models (GCMs) —the National Centre for Atmospheric Research, US (NCAR) and the Commonwealth Scientific and Industrial

Research Organization, Australia (CSIRO) models —using the A2 scenario of the IPCC Fourth Assessment Report— were used to simulate future climate. We refer to the combination of GCM model runs with A2 inputs as the NCAR and CSIRO scenarios.

Climate Change Effects on Yields

For most crops, yield declines predominate when CO2 fertilization is not included. Irrigated and rainfed wheat and irrigated rice are hard hit, and crops in South Asia are particularly negatively affected. The East Asia and Pacific region includes both China, which is temperate for the most part, and tropical Southeast Asia, so the differential effects of climate change in these two climate zones are masked. In China, some crops fare reasonably well because higher future temperatures are favourable in locations where current temperatures are at the low end of the crop's optimal range.

With the CO_2 fertilization effect included, yield declines are reduced and in many locations some yield increases occur relative to 2000. However, irrigated maize and irrigated and rainfed wheat still see substantial areas of reduced yields.

Sub-Saharan Africa sees mixed results with small declines or increases in maize yields and large negative effects on rainfed wheat. The Latin America and Caribbean region has mixed yield effects, with some crops up slightly and some down.

World Price and Production Impacts of Climate Change

The biological effects of climate change are used to alter the so-called intrinsic productivity growth rates in the IMPACT model that capture exogenous investments in productivity enhancements. The equilibrium outcomes reported below assume three scenarios for climate in 2050 – a no-climate change-scenario that assumes the 2050 climate will be identical to that around 2000 and the climate outcomes from the NCAR and CSIRO scenarios. World prices are a useful single indicator of the diverse effects of climate change on agriculture. First, even without climate change, the model results show world price increases between 2000 and 2050, a consequence of assumed population and income growth that are greater than the productivity and area growth. However, climate change makes the price increase much greater. Even with no climate change, the model estimates an increase in the price of rice of 62 percent, maize of 63 percent, soybeans of 72 percent, and wheat of 40 percent. Climate change results in additional price increases – 32 to 37 percent for rice, 52 to 55 percent for maize, 94 to 111 percent for wheat, and 11 to 14 percent for soybeans. If CO_2 fertilization is effective in farmers' fields, the 2050 price increases are smaller, with the effect varying by crop Production results in 2000 and 2050 with the three climate scenarios.

Without climate change, production of all major crops increases in developing countries. For example, in developing countries, production of rice increases by 17 percent, wheat by 76 percent and maize by 73 percent. Climate change reverses much of this increase, with the extent of the change depending on the region, crop, and climate model. For example, in South Asia, maize production increases by 15 percent with no climate change but is 9 percent below that level with the NCAR scenario and 19 percent below with the CSIRO scenario. In Sub-Saharan Africa, maize production increases by 45 percent without climate change but is 10 percent below that level with the CSIRO scenario and 7 percent lower with the NCAR scenario.

Trade in Agricultural Commodities

With no climate change, developed-country net exports increase from 83.4 million mt to 105.8 million mt between 2000 and 2050, an increase of 27 percent. Developing-country net imports mirror this change. With the NCAR results and no CO_2 fertilization, developed-country net exports increase slightly (0.9 million mt) over no climate change. With the drier CSIRO scenario, on the other hand, developed-country net exports increase by 39.9 million mt. Regional results show important differences in the effects of climate change on trade and the differential effects of the three scenarios. For example, South Asia is a small net exporter in 2000 and becomes a net importer of cereals in 2050 with no climate change. Both climate change scenarios results in substantial increases in South Asian net imports relative to no climate change. The East Asia and Pacific region is a net importing region in 2000 and imports grow substantially with no climate change. Depending on climate change scenario, this region either has slightly less net imports than with the no-climate-change scenario or becomes a net exporter. In Latin America and the Caribbean, the 2050 no-climate-change scenario is increased imports relative to 2000 but the CSIRO and NCAR climate scenarios result in smaller net imports in 2050 than in 2000.

The effects of climate change on trade flow values are even more dramatic than on production because of climate change effects on prices. Without climate change, the value of developing country net imports of cereals in 2050 is 114 percent greater than in 2000. With the wetter NCAR scenario, 2050 net imports value is 262 percent greater than in 2000; with the drier CSIRO scenario it is 361 percent greater. The climate scenario differences in trade flows are driven by geographical differences in production effects. For example, without climate change, 2050 developed country production of maize increases by 207.2 million mt (an increase of 70 percent); in developing countries, maize production increases by 234.9 million mt (73 percent). With both CSIRO and NCAR scenarios, developed country production increases more, while developing country production increases less, but the magnitudes of these changes are much greater with CSIRO than with NCAR. The result is much greater net exports of maize (and other major rainfed crops) from developed countries with CSIRO than with NCAR. Similar differences exist for wheat, where the climate change effects on yield are much more dramatic in developing countries than in developed countries.

O Trade Flow Changes Compensate For Climate Change

As mentioned above earlier literature on the climate change effects on food availability were relatively sanguine as changes in trade flows and CO_2 fertilization offset productivity effects. Our results suggest that these conclusions were too optimistic. Changes in production result in changes in per capita calorie availability. To assess the welfare effects of these changes we use a statistical relationship estimated by Smith and Haddad that relates child malnutrition to calorie availability, maternal education, access to clean drinking water and the ratio of female to male life expectancy at birth. All variables other than calorie availability are assumed to remain constant. With no climate change, only Sub-Saharan Africa would experience an increase in the number of malnourished children between 2000 and 2050 as rapid population growth offsets a declining share of malnourished children. All other parts of the developing world would experience relatively large declines in the number of malnourished children due to rapid income and agricultural productivity growth.

Climate change eliminates much of the improvement in child malnourishment levels that would occur with no climate change. For example, in East Asia and the Pacific, instead of 10 million malnourished children in 2050, the number increases to more than 14 million malnourished children under both scenarios. In South Asia, instead of 52 million malnourished children in 2050, there would be more than 58 million. In Sub-Saharan Africa, climate change is expected to increase the number by more than 11 million children. If CO_2 fertilization is in fact effective in farmers' fields, the negative effect of climate change on child malnutrition is reduced somewhat.

New IATP Report Addresses Water Governance in the 21st Century

For well over a decade, IATP has advocated for alternatives to the current water governance regime that privileges profit over people, communities and ecosystems. In advocating against neoliberal approaches to solving water crises, we have argued for the promotion of the right to water and the right to food, for the precautionary principle and for the need to respect our common but differentiated responsibility to protect our commons.

As the United Nations celebrates World Water Day, and as many organizations at the World Social Forum celebrate Water Justice Day, we offer *Water Governance in the 21st Century: Lessons from Water Trading in the U.S. and Australia,* a new paper that looks at the possibilities for water governance based on cooperation rather than competition. We look at the experiences of water trading in Australia and North America for relevant lessons to help chart a path for just and sustainable water governance in 21st century.

As water insecurities increase globally, there is an increasing emphasis on demand-management approaches, which for the most part emphasize market mechanisms as a means to ensure water security for all. Water trading is one of the market based mechanisms that helps transfer water from one user to another. It involves buying and selling water rights (which are permanent access entitlements), or water allocation entitlements (which are seasonal and temporary). This process results in the re-allocation of water among competing uses by facilitating the transfer of water from low-valued to higher-valued uses. This approach is gaining ground as climate uncertainties grow, as corporations want to control water for their value chain and as scarcity conditions give rise to the idea of water primarily as an economic good.

The Climate Change - Agriculture Conundrum

The risks associated with climate change lie in the interaction of several systems with many variables that must be collectively considered. Agriculture (including crop agriculture, animal husbandry, forestry and fisheries) can be defined as one of the systems, and climate the other. If these systems are treated independently, this would lead to an approach which is too fragmentary. The issue is more global. It is now held as likely that human activities can affect climate, one of the components of the environment. Climate in turn affects agriculture, the source of all food consumed by human beings and domestic animals. It must be further considered that not only climate may be changing, but that human societies and agriculture develop trends and constraints of their own which climate change impact studies must take into consideration.

In general, global food production has been growing faster than human population. There are, however, marked disparities between continents. For example, in Africa local production of cereals cannot keep pace with population increase, and production of root and tuber crops is growing faster than that of the nutritionally more valuable cereals.

In contrast, arable land growth lags behind population growth, which indicates some intensification of production. In Asia, the upper limit of available land has been reached in several countries, resulting in very high cropping intensities and a dominant role for irrigation. In Latin America, the increase in arable land is achieved only at a high ecological cost (especially deforestation) which may have direct relevance to climate change.

According to a recent FAO prospective study, 'the rate of growth of agricultural land will be further reduced during the next two decades. The pressures on *fresh* water *resources,* however, will be considerable, as will be those on the environment arising from the intensification of land use'.

A significant reduction in world population growth rates is foreseen from 1.8% to 1.4%, roughly equivalent to an increase of the population doubling time from 40 to 50 years. The projected continuation of the high population growth in Africa can be related to slow economic development during the coming two decades, since in general a reduction of poverty precedes reduction of population growth.

AT-2010 also lists the following significant trends for the near future:

- world production of cereals will continue to grow, but not in per capita terms;
- export cereals will undergo a modest growth in demand;
- the livestock sector in developing countries will continue to grow;
- root crops, tubers and plantains will retain their importance;
- oil crops will undergo rapid growth in developing countries; and, most importantly,
- many developing countries will become net agricultural importers.

As indicated above for population increase, large differences exist in the rate of agricultural development among the developing continents, while differences among the developed countries tend to level out.

It is also likely that hunger and under-nutrition, which currently affects 800 million people (about 20% of the 4 000 million inhabitants of developing countries), will be reduced, but large pockets of malnutrition will persist. Pressure on environmental resources will continue to build up.

The Climate 'Complex'

Climate constitutes a complex of inter-related variables. On average, through a set of regulatory mechanisms, a smooth change in one variable triggers smooth changes in most others. With the exception of possible qualitative and abrupt variations, which will be mentioned below, such inter-relations are independent of atmospheric carbon dioxide (CO_2). The latter and other greenhouse gases play a part largely through their effect on the radiation balance of the atmosphere. There is only a weak link between such factors as cloudiness and wind. Temperature, evaporation and rain are strongly correlated, which illustrates the likely intensification of the hydrological cycle. Combined with the projected pressure on land and water use, competition for land and water will certainly become a key social and political issue. Climate variability is likely to increase under global warming, both in absolute and in relative terms. This is linked with thresholds which affect the occurrence of many meteorological phenomena. For instance, tropical cyclones are 'fed' by water vapour evaporating from oceans at a temperature above 26 or 27°C. Therefore, higher average sea surface temperatures are bound to result in a higher frequency of tropical cyclones.

The rate of change itself is extremely important. For example, recent work on the saw-tooth temperature changes in the past as observed in the Arctic, raise concern that changes may occur abruptly, with average temperatures changing by 10 or 12°C in just a matter of decades. The mechanism of such changes is not clear as yet, but seems to involve the mechanical stability of ice sheets, and sudden changes in the hydrological cycle and the Atlantic conveyor current.

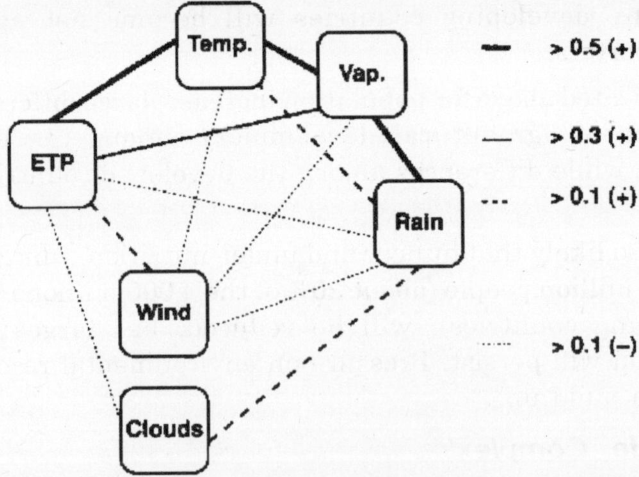

Figure: *Some relationships between major climate variables (average temperature,* Temp.; *water vapour pressure,* Vap.; *rainfall,* Rain; *wind speed,* Wind; *cloudiness,* Clouds; *and evapo-transpiration potential,* ETP*). Solid and dotted lines indicate positive and negative correlations, respectively. The strength of the correlation decreases from double heavy lines to thin single lines. Computed from annual averages of 3 263 land stations, mostly from developing countries*

Such changes would, of course, be associated with dramatic changes in the distribution and quantities of ocean products, and cause havoc to established national fishery activities. They would also make adaptation to climate change, together with most agricultural planning, extremely difficult.

Agricultural Greenhouse Gases

In addition to water vapour, important greenhouse gases are carbon dioxide (CO_2), methane (CH_4), nitrous oxide (N_2O), tropospheric ozone (O_3) and chlorofluorocarbons (CFCs). The basic characteristics of the first three gases are given. The degree to which these greenhouse gases stem from agricultural sources is also given. The exact ratio between these land-use related emissions and those from natural ecosystems (swamps, tundra), fossil fuels (coal, oil, gas) and geological sources (volcanoes) are a bone of contention among industrialized, oil producing and developing countries. Moral rights and duties in relation to the international flow of development and conservation funds are involved. Therefore, better estimates based on exact measurements of the net greenhouse gas emissions from agricultural practices in developing countries are urgently required to assess responsibilities properly. The required reductions of emissions to achieve stabilization of atmospheric concentrations of current levels are believed to be >60% for CO_2 15-20% for CH_4 and 70-80% for N_2O. This implies the need for up-to-date and complete information on land cover and land uses per country: not only the kind of crops grown but also their intensity, their rotation and the amount of inputs (energy, fertilizers).

Table: An overview of agricultural greenhouse gases with the trends as currently envisaged. ppmv and ppbv stand for parts per 10^6 and parts per 10^9, respectively, by volume.

	CO_2	CH_4	N_2O
Atmospheric lifetime (yr)	120	14.5	120
Direct GWP [1]	1	24.5 [2]	320
Pre-industrial concentration [3]	280 ppmv	0.8 ppmv	288 ppbv
Present-day levels	360 ppmv	1.72 ppmv	310 ppbv
Current annual increase (%)	0.5	0.9	0.25
Major agricultural sources [4]	deforestation	- wetland rice	- synthetic N fertilizers
		- ruminants	- animal excreta
		- biomass burning	- biological N fixation
Percentage of global source stemming from agriculture	30	40	25
Predicted change 1990-2020	-	+	+

1. GWP, Global Warming Potential, is the direct warming effect in relation to CO_2 at a time horizon of 100 years.

2. Includes indirect effects through chemistry.
3. About year 1750-1800.
4. Activities responsible for emissions are projected to increase by: rice (+ 10%); ruminant population (+30%); synthetic fertilizer use (+20%): animal excreta and biological N fixation increasing but rate not specified.

The loss of 'natural land', including tropical forests to agriculture, grazing, logging and urbanization, may continue, though at a slower pace. Many earlier estimates are now regarded with suspicion, as the borderline between crop agriculture, forest and cattle agriculture appears to be a fuzzy one.

Finally, both recent trends of fertilizer use and *AT-2010* projections indicate major increases in consumption. Even with appropriate measures to optimize fertilizer use, it is likely that N_2O losses from fertilizers will continue to increase.

Ecological and Indirect Climate Effects

In qualitative terms, many indirect effects of climate change on agriculture can be conjectured.

Most of them are estimated to be negative and they catch most of the attention of the media. These effects include:

- the overall predictability of weather and climate would decrease, making the day-to-day and medium-term planning of farm operations more difficult;
- loss of biodiversity from some of the most fragile environments, such as tropical forests and mangroves;
- sea-level rise (40 cm in the coming 100 years) would submerge some valuable coastal agricultural land;
- the incidence of diseases and pests, especially alien ones, could increase;
- present (agro) ecological zones could shift in some cases over hundreds of kilometres horizontally, and hundreds of metres altitudinally, with the hazard that some plants, especially trees, and animal species cannot follow in time, and that farming systems cannot adjust themselves in time;
- higher temperatures would allow seasonally longer plant growth and crop growing in cool and mountainous areas, allowing in some cases increased cropping and production. In contrast, in already warm areas climate change can cause reduced productivity;

- the current imbalance of food production between cool and temperate regions and tropical and subtropical regions could worsen.

Table: *Growth rates between 1961 and 1990 in agricultural sectors responsible for greenhouse gas emissions. Europe and Asia do not include the former USSR. Domestic ruminant numbers were computed as the sum of cattle, sheep, goats, camels and buffaloes*

Continent	*1961-1990 exponential growth rate (%)*			
	Ruminant numbers	*Forested area*	*Rice area*	*Fertilizer consumption*
Africa	1.29	-0.43	2.23	6.21
N and C America	-0.07	-0.02	1.50	3.29
S. America	1.29	-0.49	1.65	9.15
Asia	1.18	-0.59	0.62	9.54
Europe	0.33	0.25	0.78	2.75
Oceania	0.07	-1.15	5.88	1.25
World	**0.90**	**-0.26**	**0.74**	**5.35**

Plant Physiological Direct Effects

The greenhouse gases CH_4, N_2O and chlorofluorocarbons (CFCs) have no known direct effects on plant physiological processes. They only change global temperature and are therefore not discussed further. Instead, concentration should be on the effects of increased CO_2 tropospheric O_3, increased UV-B through depleted stratospheric ozone, increased temperatures and the associated intensification of the hydrological cycle.

Carbon Dioxide

The CO_2 fertilization Effect: CO_2 is an essential plant 'nutrient', in addition to light, suitable temperature, water and chemical elements such as N, P and K, and it is currently in short supply.

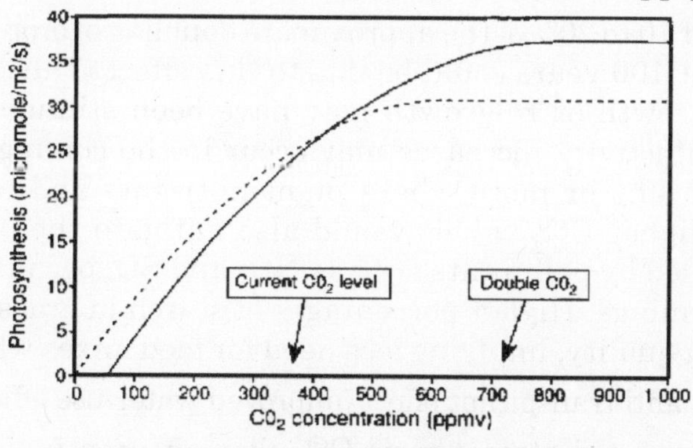

Figure: *Schematic effect of CO_2 concentrations on C_3 and C_4 plants. The main mechanism of CO_2 fertilization is that it depresses photo-respiration, more so in C_3 than in C_4 plants*

Higher concentrations of atmospheric CO_2 due to increased use of fossil fuels, deforestation and biomass burning, can have a positive influence on photosynthesis; under optimal growing conditions of light, temperature, nutrient and moisture supply, biomass production can increase, especially of plants with C3 photo-synthetic metabolism, above and even more below ground.

The Major Agricultural Crops and the Three Photosynthetic Pathways

Plants are classified as C_3, C_4 or CAM according to the products formed in the initial phases of photosynthesis.

C_3 species respond more to increased CO_2; C_4 species respond better than C_3 plants to higher temperature and their water-use efficiency increases more than for C_3 plants. There are some indications that enhancements can decline over time ('down-regulation').

C_3 plants: cotton, rice, wheat, barley, soybeans, sunflower, potatoes, most leguminous and woody plants, most horticultural crops and many weeds

C_4 plants: maize, sorghum, sugar cane, millets, halophytes (i.e., salt-tolerant plants) and many tall tropical grasses, pasture, forage and weed species

CAM plants (Crassulacean Acid Metabolism, an optional C_3 or C_4 pathway of photosynthesis, depending on conditions): cassava, pineapple, opuntia, onions, castor

A total of 10 to 20% of the approximate doubling of crop productivity over the past 100 years could be due to this effect (Tans *et al.,* 1990) and forest growth or re-growth may have been stimulated as well. Further productivity increases may occur in the coming century, in the order of 30% or more where plant nutrients and moisture are adequate. Higher CO_2 values would also mitigate the plant growth damage caused by pollutants such as NO_x and SO_2 because of smaller stomatal openings. Higher percentages of starch in grasses improves their feeding quality, implying less need for feed mixes when silaging.

The CO_2 anti-transpirant effect (improved water-use efficiency WUE)

With increased atmospheric CO_2 the consumptive use of water becomes more efficient because of reduced transpiration. This is induced by a contraction of plant stomata and/or a decrease in the number of stomata per unit leaf area. This restricts the escape of water vapour from the leaf more than it restricts photosynthesis.

With the same amount of available water, there could be more leaf area and biomass production by crops and natural vegetation. Plants could survive in areas hitherto too dry for their growth.

Ultraviolet Radiation

Increased ultraviolet radiation (UV-B, between 280 and 320 nanometres), due to depletion of the stratospheric ozone layer, mainly in the Antarctic region, may negatively affect terrestrial and aquatic photosynthesis and animal health.

Over the last decade, a decrease of stratospheric ozone was observed at all latitudes (about 10% in winter, 0% during summer and intermediate values during spring and autumn). However, the 'Biological Action Factor' of UV-B can vary over several orders of magnitude with even slight changes in the amount and wavelength of UV-B.

The subject is treated in detail elsewhere in this book. In particular it can be noted that:

- there are damaging effects of increasing UV-B on crops, animals and plankton growth. It has been reported that UV-B affects the ability of plankton organisms to control their vertical movements and to adjust to light levels;
- reductions in yield of up to 10% have been measured at experimentally very high UV-B values, and would be particularly effective in plants where the CO_2 fertilization effect is strongest. On the other hand, UV-B increase could increase the amount of plant internal compounds that act against pests.

Tropospheric Ozone

Tropospheric ozone originates about half from photochemical reactions involving nitrogen oxides (NO_x), methane or carbon monoxide, and half by downward movement of stratospheric ozone.

High ozone concentrations have toxic effects on both plant and animal life. It is likely that ozone, in conjunction with other photo-oxidants, is contributing towards the 'new type of forest damage' observed in Europe and the United States.

In the tropics, tropospheric ozone concentrations are generally lower than at northern mid-latitudes. However, this does not apply to periods when biomass burning releases precursor substances for the photochemical formation of ozone.

Temperature

In general, higher temperatures are associated with higher radiation and higher water use. It is relatively difficult to separate the physiological effects (at the level of plants and plant organs) of temperatures from the ecological ones (at the level of the field or of the region). There are both positive and negative impacts at the two levels, and only crop- and site-specific simulation can assess the global 'net' effect of temperature increases. It is generally agreed that:

- rising temperatures - now estimated to be 0.2°C per decade, or 1 °C by 2040 with smallest increases in the tropics - would diminish the yields of some crops, especially if night temperatures are increased (the temperature increase since the mid-1940s is mainly due to increasing night-time temperatures, while CO_2-induced warming would result in an almost equally large rise in minimum and maximum temperatures;
- higher temperatures could have a positive effect on growth of plants of the CAM type. They would also strengthen the CO_2 fertilization effect and the CO_2 antitranspirant effect of C_3 and C_4 plants unless plants get overheated;
- higher night temperature may increase dark respiration of plants, diminishing net biomass production;
- higher cold-season temperatures may lead to earlier ripening of annual crops, diminishing yield per crop, but would allow locally for the growth of more crops per year due to lengthening of the growing season. Winter kill of pests is likely to be reduced at high latitudes, resulting in greater crop losses and higher need for pest control;
- higher temperatures will allow for more plant growth at high latitudes and altitudes.

Combined Effects and Some Uncertainties

The changes in CO_2 tropospheric ozone and increased UV-B do not necessarily occur simultaneously: CO_2 increase is worldwide, but with a strong seasonality in middle and higher latitudes; significant increase of UV-B is largely limited to sub-polar regions (and mainly during the northern hemisphere winter months); high near-surface O_3 levels are restricted to the neighbourhood of major cities, airports, etc.

Note that increased biomass could even be associated with a decreased yield of grain (or sugar, oil, etc.) if one of the consequences of increased CO_2 will be a redistribution of biomass among plant organs (there are indications of relative increases of root growth).

The Hydrological Cycle and Soils

Even a slight increase in surface temperatures will affect evaporation, atmospheric moisture and precipitation. While it is generally agreed that rainfall will increase (by an estimated 10 to 15%), two aspects have to be elucidated: how will rainfall intensities be affected, and what are the details of spatial changes. This is still largely a matter of discussion among experts.

Based on palaeoclimatic analogies, certain authors predict more favourable rainfall conditions in the present-day Sahel. If the increase in precipitation should be associated with increased rainfall intensities, then the quality and quantity of soil and water resources would decline, for instance through increased runoff and erosion, increased land degradation processes, and a higher frequency of floods and possibly droughts. However:

- the extra precipitation on land, if indeed including present sub-humid to semi-arid areas, will increase plant growth in these areas, leading to an improved protection of the land surface and increased rainfed agricultural production; in already humid areas the extra rainfall may, however, impair adequate crop drying and storage;
- the extra precipitation predicted to occur in some regions provides possibilities for off-site extra storage in rivers, lakes and artificial reservoirs (on-farm or at sub-catchment level) for the benefit of improved rural water supply and expanded or more intensive irrigated agriculture and inland fisheries:
- the effects on water resources and water apportioning of international river and lake basins can be very substantial, with political overtones.

Some mechanisms likely to affect biomass production under global change conditions. Note that the ratio between economic yield (e.g., grain, fibre) and biomass may change relative to current conditions

ETP: Evapotranspiration potential

WHC: Soil water holding capacity

ETA: Actual evapotranspiration

OM: Organic matter

WUE: Water-use efficiency

LAI: Leaf area index

The heavy line indicates a hypothetical link between increased humidity and cloudiness. The greatest risks are often estimated to be associated with increased soil loss through erosion.

Soils, as a medium for plant growth, would be affected in several other ways:

- increased temperatures may lead to more decomposition of soil organic matter;
- increased plant growth due to the CO_2 fertilization effect may cause other plant nutrients such as N and P to become in short supply; however, CO_2 increase would stimulate mycorrhizal activity (making soil phosphorus more easily available), and also biological nitrogen fixation (whether or not symbiotic). Through increased root growth there would be extra weathering of the substratum, hence a fresh supply of potassium and micronutrients;
- the CO_2 fertilization effect would produce more litter of higher C/N ratio, hence more organic matter for incorporation into the soil as humus; litter with high C/N decomposes slowly and this can act as a negative feedback on nutrient availability;
- the 'CO_2 anti-transpirant' effect would stimulate plant growth in dryland areas, and more soil protection against erosion and lower topsoil temperatures, leading to an 'anti-desertification effect'.

Water Resources

Water resources are sources of water that are useful or potentially useful. Uses of water include agricultural, industrial, household, recreational and environmental activities. Virtually all of these human uses require fresh water. 97% of the water on the Earth is salt water. However, only three percent is fresh water; slightly over two thirds of this is frozen in glaciers and polar ice caps. The remaining unfrozen freshwater is found mainly as groundwater, with only a small fraction present above ground or in the air.

Fresh water is a renewable resource, yet the world's supply of clean, fresh water is steadily decreasing. Water demand already exceeds supply in many parts of the world and as the world population continues to rise, so too does the water demand.

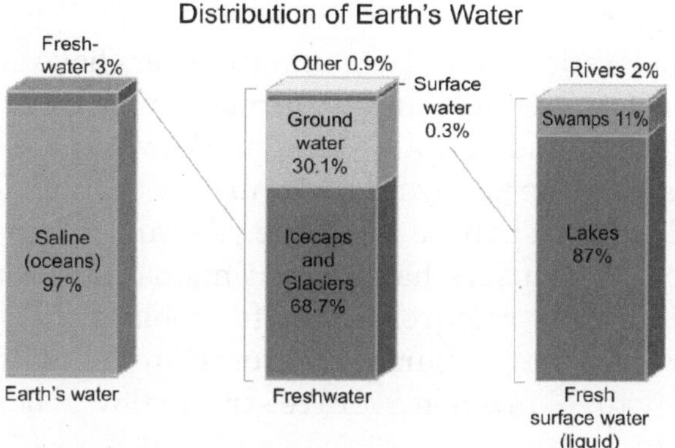

Figure: *A graphical distribution of the locations of water on Earth.*

Awareness of the global importance of preserving water for ecosystem services has only recently emerged as, during the 20th century, more than half the world's wetlands have been lost along with their valuable environmental services for Water Education. The framework for allocating water resources to water users (where such a framework exists) is known as water rights.

Sources of Fresh Water

Surface Water

Surface water is water in a river, lake or fresh water wetland. Surface water is naturally replenished by precipitation and naturally lost through discharge to the oceans, evaporation, evapotranspiration and sub-surface seepage.

Although the only natural input to any surface water system is precipitation within its watershed, the total quantity of water in that system at any given time is also dependent on many other factors. These factors include storage capacity in lakes, wetlands and artificial reservoirs, the permeability of the soil beneath these storage bodies, the runoff characteristics of the land in the watershed, the timing of the precipitation and local evaporation rates. All of these factors also affect the proportions of water loss.

Human activities can have a large and sometimes devastating impact on these factors. Humans often increase storage capacity by constructing reservoirs and decrease it by draining wetlands. Humans often increase runoff quantities and velocities by paving areas and channelizing stream flow.

The total quantity of water available at any given time is an important consideration. Some human water users have an intermittent need for water. For example, many farms require large quantities of water in the spring, and no water at all in the winter. To supply such a farm with water, a surface water system may require a large storage capacity to collect water throughout the year and release it in a short period of time. Other users have a continuous need for water, such as a power plant that requires water for cooling. To supply such a power plant with water, a surface water system only needs enough storage capacity to fill in when average stream flow is below the power plant's need.

Nevertheless, over the long term the average rate of precipitation within a watershed is the upper bound for average consumption of natural surface water from that watershed. Natural surface water can be augmented by importing surface water from another watershed through a canal or pipeline. It can also be artificially augmented from any of the other sources listed here, however in practice the quantities are negligible. Humans can also cause surface water to be "lost" (i.e. become unusable) through pollution. Brazil is the country estimated to have the largest supply of fresh water in the world, followed by Russia and Canada.

Under River Flow

Throughout the course of a river, the total volume of water transported downstream will often be a combination of the visible free water flow together with a substantial contribution flowing through sub-surface rocks and gravels that underlie the river and its floodplain called the hyporheic zone. For many rivers in large valleys, this unseen component of flow may greatly exceed the visible flow. The hyporheic zone often forms a dynamic interface between surface water and true ground-water receiving water from the ground water when aquifers are fully charged and contributing water to ground-water when ground waters are depleted. This is especially significant in karst areas where pot-holes and underground rivers are common.

Ground Water

Sub-surface water, or groundwater, is fresh water located in the pore space of soil and rocks. It is also water that is flowing within aquifers below the water table. Sometimes it is useful to make a distinction between sub-surface water that is closely associated with surface water and deep sub-surface water in an aquifer (sometimes called "fossil water").

Sub-surface water can be thought of in the same terms as surface water: inputs, outputs and storage. The critical difference is that due to its slow rate of turnover, sub-surface water storage is generally much larger compared to inputs than it is for surface water. This difference makes it easy for humans to use sub-surface water unsustainably for a long time without severe consequences.

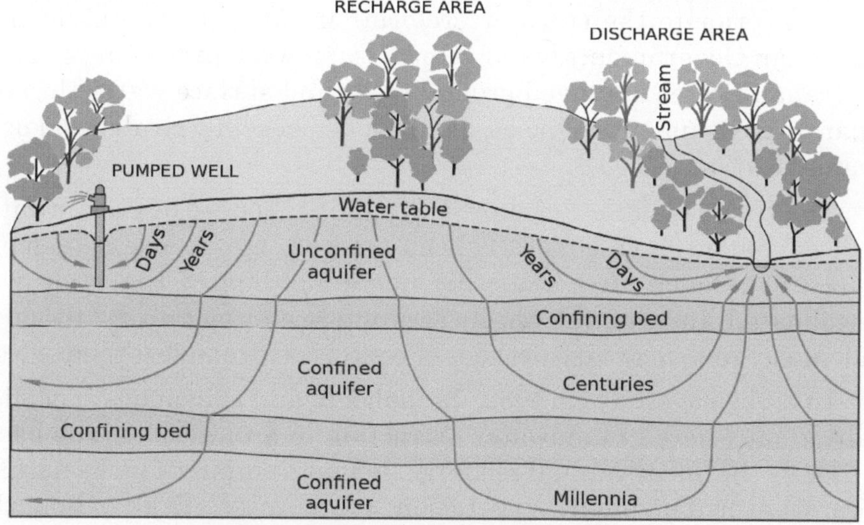

Figure: *Sub-Surface water travel time*

The natural input to sub-surface water is seepage from surface water. The natural outputs from sub-surface water are springs and seepage to the oceans.

If the surface water source is also subject to substantial evaporation, a sub-surface water source may become saline. This situation can occur naturally under endorheic bodies of water, or artificially under irrigated farmland.

In coastal areas, human use of a sub-surface water source may cause the direction of seepage to ocean to reverse which can also cause soil salinization. Humans can also cause sub-surface water to be "lost" (i.e. become unusable) through pollution. Humans can increase the input to a sub-surface water source by building reservoirs or detention ponds.

Issues

Certain problems have beset the use of groundwater around the world. Just as river waters have been over-used and polluted in many parts of the world, so too have aquifers. The big difference is that aquifers are out of sight. The other major problem is that water management agencies, when calculating the "sustainable yield" of aquifer and river water, have often counted the same water twice, once in the aquifer, and once in its connected river.

This problem, although understood for centuries, has persisted, partly through inertia within government agencies. In Australia, for example, prior to the statutory reforms initiated by the Council of Australian Governments water reform framework in the 1990s, many Australian States managed groundwater and surface water through separate government agencies, an approach beset by rivalry and poor communication.

The time lags inherent in the dynamic response of groundwater to development have generally been ignored by water management agencies, decades after scientific understanding of the issue was consolidated. In brief, the effects of groundwater overdraft (although undeniably real) may take decades or centuries to manifest themselves.

In a classic study in 1982, Bredehoeft and colleagues modelled a situation where groundwater extraction in an intermontane basin withdrew the entire annual recharge, leaving 'nothing' for the natural groundwater-dependent vegetation community. Even when the borefield was situated close to the vegetation, 30% of the original vegetation demand could still be met by the lag inherent in the system after 100 years.

By year 500 this had reduced to 0%, signalling complete death of the groundwater-dependent vegetation. The science has been available to make these calculations for decades; however water management agencies have generally ignored effects which will appear outside the rough timeframe of political elections (3 to 5 years). Marios Sophocleous argued strongly that management agencies must define and use appropriate timeframes in groundwater planning. This will mean calculating groundwater withdrawal permits based on predicted effects decades, sometimes centuries in the future. As water moves through the landscape it collects soluble salts, mainly sodium chloride. Where such water enters the atmosphere through evapotranspiration, these salts are left behind. In irrigation districts, poor drainage of soils and surface aquifers can result in water tables coming to the surface in low-lying areas. Major land degradation problems of soil salinity and waterlogging result, combined with increasing levels of salt in surface waters. As a consequence, major damage has occurred to local economies and environments.

Four important effects are worthy of brief mention. First, flood mitigation schemes, intended to protect infrastructure built on floodplains, have had the unintended consequence of reducing aquifer recharge associated with natural flooding. Second, prolonged depletion of groundwater in extensive aquifers can result in land subsidence, with associated infrastructure damage – as well as (thirdly) saline intrusion. Fourth, draining acid sulphate soils, often found in low-lying coastal plains, can result in acidification and pollution of formerly freshwater and estuarine streams.

Another cause for concern is that groundwater drawdown from over-allocated aquifers has the potential to cause severe damage to both terrestrial and aquatic ecosystems – in some cases very conspicuously but in others quite imperceptibly because of the extended period over which the damage occurs.

Overdraft

Groundwater is a highly useful and often abundant resource. However, over-use, or overdraft, can cause major problems to human users and to the environment. The most evident problem (as far as human groundwater use is concerned) is a lowering of the water table beyond the reach of existing wells. Wells must consequently be drilled deeper to reach the groundwater; in some places (e.g., California, Texas and India) the water table has dropped hundreds of feet because of extensive well pumping. In the Punjab region of India, for example, groundwater levels have dropped 10 metres since 1979, and the rate of depletion is accelerating. A lowered water table may, in turn, cause other problems such as groundwater-related subsidence and saltwater intrusion.

Groundwater is also ecologically important. The importance of groundwater to ecosystems is often overlooked, even by freshwater biologists and ecologists. Groundwaters sustain rivers, wetlands and lakes, as well as subterranean ecosystems within karst or alluvial aquifers.

Not all ecosystems need groundwater, of course. Some terrestrial ecosystems – for example, those of the open deserts and similar arid environments – exist on irregular rainfall and the moisture it delivers to the soil, supplemented by moisture in the air.

While there are other terrestrial ecosystems in more hospitable environments where groundwater plays no central role, groundwater is in fact fundamental to many of the world's major ecosystems. Water flows between groundwaters and surface waters. Most rivers, lakes and wetlands are fed by, and (at other places or times) feed groundwater, to varying degrees.

Groundwater feeds soil moisture through percolation, and many terrestrial vegetation communities depend directly on either groundwater or the percolated soil moisture above the aquifer for at least part of each year. Hyporheic zones (the mixing zone of streamwater and groundwater) and riparian zones are examples of ecotones largely or totally dependent on groundwater.

A Possible Solution to Over-use of Groundwater in India

As ageing large-scale surface irrigation schemes have become increasingly inefficient, and farmers have begun growing a wider range of crops requiring water on demand, the number of groundwater wells in India has exploded. In 1960, there were fewer than 100,000 such wells; by 2006 the figure had risen to nearly 12 million. In India, a possible solution to over-use of groundwater is emerging, known as 'groundwater recharge'.

It involves capturing rainwater that would otherwise run-off, and using it to refill aquifers. Since 2000, the International Water Management Institute (IWMI) has been working with the Indian authorities to help improve the availability of water for agriculture in India. In 2006, India's finance minister invited IWMI to submit policy recommendations based on its research on groundwater depletion.

One of the key recommendations was to instigate a programme of recharging groundwater across the 65 per cent of India that has hard-rock aquifers. As a result, the Indian government allocated Rs 1800 crore (US$400million) to fund dug-well recharge projects (a dug-well is a wide, shallow well, often lined with concrete) in 100 districts within seven states where water stored in hard-rock aquifers has been over-exploited. These geological formations have a much lower capacity to store rainwater than alluvial areas with porous sand or clay rocks, hence being given priority. The money is sufficient to fund seven million structures to be installed on dug-wells to divert monsoon runoff. The structures include a de-siltation chamber, plus pipes to collect surplus rainwater and divert de-silted water from the chamber to the well. As of the end of November 2009, funds amounting to Rs. 216.98 crore (including Rs. 199.98 crore as subsidy to beneficiaries and 17 crore for IEC/Capacity Building activities) had been released to the concerned states. Subsidies had been released to 566,637 beneficiaries.

Subsidence

Subsidence occurs when too much water is pumped out from underground, deflating the space below the above-surface, and thus causing the ground to actually collapse. The result can look like craters on plots of land.

This occurs because in its natural equilibrium state, the hydraulic pressure of groundwater in the pore spaces of the aquifer and the aquitard supports some of the weight of the overlying sediments. When groundwater is removed from aquifers by excessive pumping, pore pressures in the aquifer drop and compression of the aquifer may occur.

This compression may be partially recoverable if pressures rebound, but much of it is not. When the aquifer gets compressed it may cause land subsidence, a drop in the ground surface.

The city of New Orleans, Louisiana, is actually below sea level today, and its subsidence is partly caused by removal of groundwater from the various aquifer/aquitard systems beneath it. In the first half of the 20th century, the city of San Jose, California, dropped 13 feet from land subsidence caused by overpumping; this subsidence has been halted with improved groundwater management.

Seawater Intrusion

Generally, in very humid or undeveloped regions, the shape of the water table mimics the slope of the surface. The recharge zone of an aquifer near the seacoast is likely to be inland, often at considerable distance.

In these coastal areas, a lowered water table may induce sea water to reverse the flow towards the land. Sea water moving inland is called a saltwater intrusion. Alternatively, salt from mineral beds may leach into the groundwater of its own accord.

Mining

Sometimes the water movement from the recharge zone to the place where it is withdrawn may take centuries. When the usage of water is greater than the recharge, it is referred to as *mining* water (the water is often called fossil water because of its geologic age). Under those circumstances it is not a renewable resource.

Desalination

Desalination is an artificial process by which saline water (generally sea water) is converted to fresh water. The most common desalination processes are distillation and reverse osmosis. Desalination is currently expensive compared to most alternative sources of water, and only a very small fraction of total human use is satisfied by desalination. It is only economically practical for high-valued uses (such as household and industrial uses) in arid areas. The most extensive use is in the Persian Gulf.

Frozen .Water

Several schemes have been proposed to make use of icebergs as a water source, however to date this has only been done for novelty purposes. Glacier runoff is considered to be surface water.

The Himalayas, which are often called "The Roof of the World", contain some of the most extensive and rough high altitude areas on Earth as well as the greatest area of glaciers and permafrost outside of the poles. Ten of Asia's largest rivers flow from there, and more than a billion people's livelihoods depend on them. To complicate matters, temperatures are rising more rapidly here than the global average. In Nepal the temperature has risen with 0.6 degree over the last decade, whereas the global warming has been around 0.7 over the last hundred years.

Agricultural

It is estimated that 69% of worldwide water use is for irrigation, with 15-35% of irrigation withdrawals being unsustainable. It takes around 2,000 - 3,000 litres of water to produce enough food to satisfy one person's daily dietary need. This is a considerable amount, when compared to that required for drinking, which is between two and five litres. To produce food for the now over 7 billion people who inhabit the planet today requires the water that would fill a canal ten metres deep, 100 metres wide and 2100 kilometres long.

Increasing Water Scarcity

Fifty years ago, the common perception was that water was an infinite resource. At this time, there were fewer than half the current number of people on the planet. People were not as wealthy as today, consumed fewer calories and ate less meat, so less water was needed to produce their food. They required a third of the volume of water we presently take from rivers. Today, the competition for water resources is much more intense.

This is because there are now seven billion people on the planet, their consumption of water-thirsty meat and vegetables is rising, and there is increasing competition for water from industry, urbanisation biofuel crops, and water reliant food items. In future, even more water will be needed to produce food because the Earth's population is forecast to rise to 9 billion by 2050. An additional 2.5 or 3 billion people, choosing to eat fewer cereals and more meat and vegetables could add an additional five million kilometres to the virtual canal mentioned above.

An assessment of water management in agriculture was conducted in 2007 by the International Water Management Institute in Sri Lanka to see if the world had sufficient water to provide food for its growing population. It assessed the current availability of water for agriculture on a global scale and mapped out locations suffering from water scarcity. It found that a fifth of the world's people, more than 1.2 billion, live in areas of physical water scarcity, where there is not enough water to meet all demands. One third of the worlds population does not have access to clean drinking water, which is more than 2.3 billion people. A further 1.6 billion people live in areas experiencing economic water scarcity, where the lack of investment in water or insufficient human capacity make it impossible for authorities to satisfy the demand for water. The report found that it would be possible to produce the food required in future, but that continuation of today's food production and environmental trends would lead to crises in many parts of the world. To avoid a global water crisis, farmers will have to strive to increase productivity to meet growing demands for food, while industry and cities find ways to use water more efficiently.

In some areas of the world irrigation is necessary to grow any crop at all, in other areas it permits more profitable crops to be grown or enhances crop yield. Various irrigation methods involve different trade-offs between crop yield, water consumption and capital cost of equipment and structures.

Irrigation methods such as furrow and overhead sprinkler irrigation are usually less expensive but are also typically less efficient, because much of the water evaporates, runs off or drains below the root zone. Other irrigation methods considered to be more efficient include drip or trickle irrigation, surge irrigation, and some types of sprinkler systems where the sprinklers are operated near ground level. These types of systems, while more expensive, usually offer greater potential to minimize runoff, drainage and evaporation. Any system that is improperly managed can be wasteful, all methods have the potential for high efficiencies under suitable conditions, appropriate irrigation timing and management. Some issues that are often insufficiently considered are salinization of sub-surface water and contaminant accumulation leading to water quality declines.

As global populations grow, and as demand for food increases in a world with a fixed water supply, there are efforts under way to learn how to produce more food with less water, through improvements in irrigation methods and technologies, agricultural water management, crop types, and water monitoring. Aquaculture is a small but growing agricultural use of water. Freshwater commercial fisheries may also be considered as agricultural uses of water, but have generally been assigned a lower priority than irrigation.

Industrial

It is estimated that 22% of worldwide water is used in industry. Major industrial users include hydroelectric dams, thermoelectric power plants, which use water for cooling, ore and oil refineries, which use water in chemical processes, and manufacturing plants, which use water as a solvent. Water withdrawal can be very high for certain industries, but consumption is generally much lower than that of agriculture.

Water is used in renewable power generation. Hydroelectric power derives energy from the force of water flowing downhill, driving a turbine connected to a generator. This hydroelectricity is a low-cost, non-polluting, renewable energy source. Significantly, hydroelectric power can also be used for load following unlike most renewable energy sources which are intermittent. Ultimately, the energy in a hydroelectric powerplant is supplied by the sun.

Heat from the sun evaporates water, which condenses as rain in higher altitudes and flows downhill. Pumped-storage hydroelectric plants also exist, which use grid electricity to pump water uphill when demand is low, and use the stored water to produce electricity when demand is high.

Hydroelectric power plants generally require the creation of a large artificial lake. Evaporation from this lake is higher than evaporation from a river due to the larger surface area exposed to the elements, resulting in much higher water consumption. The process of driving water through the turbine and tunnels or pipes also briefly removes this water from the natural environment, creating water withdrawal. The impact of this withdrawal on wildlife varies greatly depending on the design of the powerplant.

Pressurized water is used in water blasting and water jet cutters. Also, very high pressure water guns are used for precise cutting. It works very well, is relatively safe, and is not harmful to the environment. It is also used in the cooling of machinery to prevent overheating, or prevent saw blades from overheating. This is generally a very small source of water consumption relative to other uses.Water is also used in many large scale industrial processes, such as thermoelectric power production, oil refining, fertilizer production and other chemical plant use, and natural gas extraction from shale rock. Discharge of untreated water from industrial uses is pollution. Pollution includes discharged solutes (chemical pollution) and increased water temperature (thermal pollution).

Industry requires pure water for many applications and utilizes a variety of purification techniques both in water supply and discharge. Most of this pure water is generated on site, either from natural freshwater or from municipal grey water. Industrial consumption of water is generally much lower than withdrawal, due to laws requiring industrial grey water to be treated and returned to the environment. Thermoelectric powerplants using cooling towers have high consumption, nearly equal to their withdrawal, as most of the withdrawn water is evaporated as part of the cooling process. The withdrawal, however, is lower than in once-through cooling systems.

Household

It is estimated that 8% of worldwide water use is for household purposes. These include drinking water, bathing, cooking, sanitation, and gardening. Basic household water requirements have been estimated by Peter Gleick at around 50 litres per person per day, excluding water for gardens. Drinking water is water that is of sufficiently high quality so that it can be consumed or used without risk of immediate or long term harm. Such water is commonly called potable water. In most developed countries, the water supplied to households, commerce and industry is all of drinking water standard even though only a very small proportion is actually consumed or used in food preparation.

Recreation

Recreational water use is usually a very small but growing percentage of total water use. Recreational water use is mostly tied to reservoirs. If a reservoir is kept fuller than it would otherwise be for recreation, then the water retained could be categorized as recreational usage. Release of water from a few reservoirs is also timed to enhance whitewater boating, which also could be considered a recreational usage. Other examples are anglers, water skiers, nature enthusiasts and swimmers.

Recreational usage is usually non-consumptive. Golf courses are often targeted as using excessive amounts of water, especially in drier regions. It is, however, unclear whether recreational irrigation (which would include private gardens) has a noticeable effect on water resources. This is largely due to the unavailability of reliable data. Additionally, many golf courses utilize either primarily or exclusively treated effluent water, which has little impact on potable water availability.

Some governments, including the Californian Government, have labelled golf course usage as agricultural in order to deflect environmentalists' charges of wasting water. However, using the above figures as a basis, the actual statistical effect of this reassignment is close to zero. In Arizona, an organized lobby has been established in the form of the Golf Industry Association, a group focused on educating the public on how golf impacts the environment.

Recreational usage may reduce the availability of water for other users at specific times and places. For example, water retained in a reservoir to allow boating in the late summer is not available to farmers during the spring planting season. Water released for whitewater rafting may not be available for hydroelectric generation during the time of peak electrical demand.

Environmental

Explicit environment water use is also a very small but growing percentage of total water use. Environmental water may include water stored in impoundments and released for environmental purposes (held environmental water), but more often is water retained in waterways through regulatory limits of abstraction. Environmental water usage includes watering of natural or artificial wetlands, artificial lakes intended to create wildlife habitat, fish ladders, and water releases from reservoirs timed to help fish spawn, or to restore more natural flow regimes

Like recreational usage, environmental usage is non-consumptive but may reduce the availability of water for other users at specific times and places. For example, water release from a reservoir to help fish spawn may not be available to farms upstream, and water retained in a river to maintain waterway health would not be available to water abstractors downstream.

Water Stress

The concept of water stress is relatively simple: According to the World Business Council for Sustainable Development, it applies to situations where there is not enough water for all uses, whether agricultural, industrial or domestic. Defining thresholds for stress in terms of available water per capita is more complex, however, entailing assumptions about water use and its efficiency. Nevertheless, it has been proposed that when annual per capita renewable freshwater availability is less than 1,700 cubic meters, countries begin to experience periodic or regular water stress. Below 1,000 cubic meters, water scarcity begins to hamper economic development and human health and well-being.

Population Growth

In 2000, the world population was 6.2 billion. The UN estimates that by 2050 there will be an additional 3.5 billion people with most of the growth in developing countries that already suffer water stress. Thus, water demand will increase unless there are corresponding increases in water conservation and recycling of this vital resource. In building on the data presented here by the UN, the World Bank goes on to explain that access to water for producing food will be one of the main challenges in the decades to come. Access to water will need to be balanced with the importance of managing water itself in a sustainable way while taking into account the impact of climate change, and other environmental and social variables.

Expansion of Business Activity

Business activity ranging from industrialization to services such as tourism and entertainment continues to expand rapidly. This expansion requires increased water services including both supply and sanitation, which can lead to more pressure on water resources and natural ecosystem

Rapid Urbanization

The trend towards urbanization is accelerating. Small private wells and septic tanks that work well in low-density communities are not feasible within high-density urban areas. Urbanization requires significant investment in water infrastructure in order to deliver water to individuals and to process the concentrations of wastewater – both from individuals and from business. These polluted and contaminated waters must be treated or they pose unacceptable public health risks.

In 60% of European cities with more than 100,000 people, groundwater is being used at a faster rate than it can be replenished. Even if some water remains available, it costs more and more to capture it.

Climate Change

Climate change could have significant impacts on water resources around the world because of the close connections between the climate and hydrological cycle. Rising temperatures will increase evaporation and lead to increases in precipitation, though there will be regional variations in rainfall. Overall, the global supply of freshwater will increase. Both droughts and floods may become more frequent in different regions at different times, and dramatic changes in snowfall and snow melt are expected in mountainous areas. Higher temperatures will also affect water quality in ways that are not well understood. Possible impacts include increased eutrophication. Climate change could also mean an increase in demand for farm irrigation, garden sprinklers, and perhaps even swimming pools. There is now ample evidence that increased hydrologic variability and change in climate has and will continue have a profound impact on the water sector through the hydrologic cycle, water availability, water demand, and water allocation at the global, regional, basin, and local levels.

Depletion of Aquifers

Due to the expanding human population, competition for water is growing such that many of the worlds major aquifers are becoming depleted. This is due both for direct human consumption as well as agricultural irrigation by groundwater. Millions of pumps of all sizes are currently extracting groundwater throughout the world. Irrigation in dry areas such as northern China and India is supplied by groundwater, and is being extracted at an unsustainable rate. Cities that have experienced aquifer drops between 10 to 50 metres include Mexico City, Bangkok, Manila, Beijing, Madras and Shanghai.

Pollution and Water Protection

Water pollution is one of the main concerns of the world today. The governments of numerous countries have striven to find solutions to reduce this problem. Many pollutants threaten water supplies, but the most widespread, especially in developing countries, is the discharge of raw sewage into natural waters; this method of sewage disposal is the most common method in underdeveloped countries, but also is prevalent in quasi-developed countries such as China, India and Iran. Sewage, sludge, garbage, and even toxic pollutants are all dumped into the water. Even if sewage is treated, problems still arise. Treated sewage forms sludge, which may be placed in landfills, spread out on land, incinerated or dumped at sea. In addition to sewage, nonpoint source pollution such as agricultural runoff is a significant source of pollution in some parts of the world, along with urban stormwater runoff and chemical wastes dumped by industries and governments.

Water and Conflicts

Competition for water has widely increased, and it has become more difficult to conciliate the necessities for water supply for human consumption, food production, ecosystems and other uses. Water administration is frequently involved in contradictory and complex problems. Approximately 10% of the worldwide annual runoff is used for human necessities. Several areas of the world are flooded, while others have such low precipitations that human life is almost impossible. As population and development increase, raising water demand, the possibility of problems inside a certain country or region increases, as it happens with others outside the region.

Over the past 25 years, politicians, academics and journalists have frequently predicted that disputes over water would be a source of future wars. Commonly cited quotes include: that of former Egyptian Foreign Minister and former Secretary-General of the United Nations Boutrous Ghali, who forecast, "The next war in the Middle East will be fought over water, not politics"; his successor at the UN, Kofi Annan, who in 2001 said, "Fierce competition for fresh water may well become a source of conflict and wars in the future," and the former Vice President of the World Bank, Ismail Serageldin, who said the wars of the next century will be over water unless significant changes in governance occurred. The water wars hypothesis had its roots in earlier research carried out on a small number of transboundary rivers such as the Indus, Jordan and Nile. These particular rivers became the focus because they had experienced water-related disputes. Specific events cited as evidence include Israel's bombing of Syria's attempts to divert the Jordan's headwaters, and military threats by Egypt against any country building dams in the upstream waters of the Nile. However, while some links made between conflict and water were valid, they did not necessarily represent the norm.

The only known example of an actual inter-state conflict over water took place between 2500 and 2350 BC between the Sumerian states of Lagash and Umma. Water stress has most often led to conflicts at local and regional levels. Tensions arise most often within national borders, in the downstream areas of distressed river basins. Areas such as the lower regions of China's Yellow River or the Chao Phraya River in Thailand, for example, have already been experiencing water stress for several years. Water stress can also exacerbate conflicts and political tensions which are not directly caused by water. Gradual reductions over time in the quality and/or quantity of fresh water can add to the instability of a region by depleting the health of a population, obstructing economic development, and exacerbating larger conflicts.

Shared Water Resources can Promote Collaboration

Water resources that span international boundaries, are more likely to be a source of collaboration and cooperation, than war.

Scientists working at the International Water Management Institute, in partnership with Aaron Wolf at Oregon State University, have been investigating the evidence behind water war predictions.

Their findings show that, while it is true there has been conflict related to water in a handful of international basins, in the rest of the world's approximately 300 shared basins the record has been largely positive. This is exemplified by the hundreds of treaties in place guiding equitable water use between nations sharing water resources. The institutions created by these agreements can, in fact, be one of the most important factors in ensuring cooperation rather than conflict.

The International Union for the Conservation of Nature (IUCN) published the book *Share: Managing water across boundaries*. One chapter covers the functions of trans-boundary institutions and how they can be designed to promote cooperation, overcome initial disputes and find ways of coping with the uncertainty created by climate change. It also covers how the effectiveness of such institutions can be monitored.

World Water Supply and Distribution

Food and water are two basic human needs. However, global coverage figures from 2002 indicate that, of every 10 people:

- roughly 5 have a connection to a piped water supply at home (in their dwelling, plot or yard);
- 3 make use of some other sort of improved water supply, such as a protected well or public standpipe;
- 2 are unserved;
- In addition, 4 out of every 10 people live without improved sanitation.

At Earth Summit 2002 governments approved a Plan of Action to:

- Halve by 2015 the proportion of people unable to reach or afford safe drinking water. The Global Water Supply and Sanitation Assessment 2000 Report (GWSSAR) defines "Reasonable access" to water as at least 20 litres per person per day from a source within one kilometre of the user's home.
- Halve the proportion of people without access to basic sanitation. The GWSSR defines "Basic sanitation" as private

or shared but not public disposal systems that separate waste from human contact.

As the picture shows, in 2025, water shortages will be more prevalent among poorer countries where resources are limited and population growth is rapid, such as the Middle East, Africa, and parts of Asia. By 2025, large urban and peri-urban areas will require new infrastructure to provide safe water and adequate sanitation. This suggests growing conflicts with agricultural water users, who currently consume the majority of the water used by humans.

Generally speaking the more developed countries of North America, Europe and Russia will not see a serious threat to water supply by the year 2025, not only because of their relative wealth, but more importantly their populations will be better aligned with available water resources. North Africa, the Middle East, South Africa and northern China will face very severe water shortages due to physical scarcity and a condition of overpopulation relative to their carrying capacity with respect to water supply. Most of South America, Sub-Saharan Africa, Southern China and India will face water supply shortages by 2025; for these latter regions the causes of scarcity will be economic constraints to developing safe drinking water, as well as excessive population growth.

1.6 billion people have gained access to a safe water source since 1990. The proportion of people in developing countries with access to safe water is calculated to have improved from 30 percent in 1970 to 71 percent in 1990, 79 percent in 2000 and 84 percent in 2004. This trend is projected to continue.

Economic Considerations

Water supply and sanitation require a huge amount of capital investment in infrastructure such as pipe networks, pumping stations and water treatment works. It is estimated that Organisation for Economic Co-operation and Development (OECD) nations need to invest at least USD 200 billion per year to replace aging water infrastructure to guarantee supply, reduce leakage rates and protect water quality.

International attention has focused upon the needs of the developing countries. To meet the Millennium Development Goals targets of halving the proportion of the population lacking access to safe drinking water and basic sanitation by 2015, current annual investment on the order of USD 10 to USD 15 billion would need to be roughly doubled. This does not include investments required for the maintenance of existing infrastructure.

Once infrastructure is in place, operating water supply and sanitation systems entails significant ongoing costs to cover personnel, energy, chemicals, maintenance and other expenses. The sources of money to meet these capital and operational costs are essentially either user fees, public funds or some combination of the two.

But this is where the economics of water management start to become extremely complex as they intersect with social and broader economic policy. Such policy questions are beyond the scope of this article, which has concentrated on basic information about water availability and water use. They are, nevertheless, highly relevant to understanding how critical water issues will affect business and industry in terms of both risks and opportunities.

2

The Effects of Elevated CO_2 and Temperature Change on Transpiration and Crop Water Use

The projected climatic effects of the continuously increasing concentrations of CO_2 and other radiatively active trace gasses in the atmosphere have caused concern over the last decades and increasingly attracted scientists' and policy makers' attention. The expected changes at a global level will be reflected in changed weather conditions in the growing season at regional and local levels that directly affect agriculture and natural vegetation.

The exchange of water and energy is determined both by the climate and by the gas exchange properties of the vegetation. Feedback on land cover, changed vegetation properties and local and regional climate is therefore of prime importance.

So far, the General Circulation Models (GCMs) are not able to give accurate predictions of climate change with great geographical detail. Also, feedback processes by the biosphere are still poorly represented in the models. These shortcomings limit predictions at the regional level that are needed in impact studies for agricultural production. Qualitative changes in potential production may well be estimated on the basis of the scenarios, but quantitative changes in the actual production, including the limitations caused by the changed variability of weather conditions (daily and seasonal amplitudes of temperatures, seasonal patterns in precipitation, cloudiness) remain difficult to estimate.

In addition to changes in precipitation and total water availability for irrigation, that directly affect agricultural production, changes in the pattern of water use by crop plants throughout the season may affect the outcome. Of special concern is the change in the physiological functioning of the vegetation as a consequence of the changed atmospheric composition.

Most plants react to the changed atmospheric CO_2 concentration with changed stomatal response, and not only is growth affected but also the transpiration. The complex nature of the physiological response in interaction with micrometeorological processes at the leaf and canopy level requires further attention.

This chapter gives an indication of the state of the art regarding the effects of elevated CO_2 on stomatal behaviour, transpiration, water-use efficiency and total water-use. It briefly touches upon the interaction between vegetation and atmosphere at a larger scale. Aspects related to these effects at the level characteristic for a doubling of the present atmospheric CO_2 concentration [CO_2] are discussed. Changes in atmospheric concentration will not occur overnight but are expected to lag behind relative to changes in greenhouse gas concentration. Therefore uncertainties remain not only on the expected time course of modifications and adaptations, but also on the level at which new threats and opportunities arise in future conditions.

Water Use and Water-use Efficiency

Definitions : It should be realized that various expressions for water-use efficiency are used, and that the physiological and agronomic definitions may differ.

Time-averaged water use and water-use efficiency (seasonal or periodic whole-plant transpiration rate (WPTR), and water-use efficiency (WUE)) are often confounded with instantaneous transpiration rate and transpiration efficiency (ITE). The latter expression is best described as the ratio between the instantaneous CO_2 fixation (actual net photosynthesis) and water loss by transpiration (both measured with a leaf or plant chamber). Water-use efficiency is the ratio of the net gain in dry matter over a given period, divided by the water loss (from the vegetation alone or from soil and vegetation together) over the same period. Mostly the ITE, normally only measured during the light period, is higher than WUE, both because of variations in ITE over day and season and because of carbon losses by respiration in the dark. WUE seems to be the more relevant parameter to the question of impact on agricultural production.

Interactions with Experimental Conditions

Most available data on water use and water-use efficiency have been collected from single plants grown in pots or hydroponically, and using porometer measurements supplemented with periodic weight determinations. Values for water-use efficiency and total water use in (semi-) field conditions over the full season, or an extended period, are scarce. Moreover, a complication for the correct estimation of the effect of elevated [CO_2] and temperature changes on transpiration is that almost all experiments in this field have been performed in environmentally controlled and generally well mixed and ventilated experimental set-ups (enclosures or Open-Top Chambers (OTCs)), where the indirect effects may not show up so prominently as in a real and outside future climate. Similar interpretation problems have been met in the evaluation of air pollution research.

Response at the Level of Stomata

Stomatal Density : The plant leaves lose water primarily by evaporation through the stomata. The stomatal density depends upon plant species, and can be related to the plant-ecotype (between 300 and 800 stomata/mm. Woodward (1987) correlated the decrease in the stomatal density over time, observed in herbarium leaves collected over the last centuries, with the rising CO_2 concentrations and concluded from the shift in δ ^{13}C that the water-use efficiency concomitantly has improved. The nitrogen content in the leaves had dropped, in line with most data from elevated CO_2 experiments. Experimentally, an increase in [CO_2] up to about 310 μ l/l decreased the stomatal density, but sometimes no effect is found above this [CO_2]. This matter is still under dispute although such a correlation has also been confirmed for palaeo-records. Among species large differences in response of stomatal density to elevated [CO_2] seem to exist. Experiments with a range of CO_2 concentrations (160-900 μ l/ l) showed an increase of stomatal density of rice and bean leaves, with a differential effect at abaxial (increasing) and adaxial sides. At subambient CO_2 concentrations stomatal density dropped. This is in contrast with the findings of Oberbauer *et al.* (1985) for tropical trees. The relative effect of changes in stomatal density and decreasing stomatal aperture at elevated [CO_2] for the water relations, however, have not been evaluated. A gradual change in [CO_2] over the next century may lead to a natural selection that favours cultivars having a lower stomatal density, especially for water-limited growth conditions. It should, however, be realized that other environmental factors like salt stress can also modify stomatal density (Rozema *et al.,* 1991a). Whatever the net effect, the resulting stomatal conductance is primarily determined by stomatal functioning, and much less by density.

Stomatal Functioning

In the pathway from the stomatal cavity to the leaf surface, and from ambient air to the photosynthetic machinery in the mesophyll, the stomata are a major resistance for gas transport between the leaf and the surrounding air.

A change in gas exchange resistance of the stomatal pores therefore affects the entrance of CO_2 and even more the exit of water vapour. The opening status of the stomata is a compromise between water loss and uptake of CO_2 from ambient air. In line with this proposition, stomatal response to elevated concentrations of CO_2 (C_a) in general is reflected in partial stomatal closure.

The mechanism behind this stomatal closure is not yet clear. The observations are in line with the notion that plants tend to regulate the internal CO_2 concentration (C_i) in the substomatal cavity such that for a given water vapour deficit there is a constant ratio (C_i/C_a) with the atmospheric concentration. Such a regulation would directly lead to the partial closure at elevated CO_2, as observed in many studies using porometers.

Jackson *et al.* (1994) measured photosynthesis and water relations of native grassland species and calculated C_i/C_a for C_3 and C_4 plants. They confirmed the conservation of the value, with only a small (not significant) tendency to rise with increasing [CO_2]. The rate of photosynthesis and therefore the required supply of carbon dioxide is directly coupled to light intensity.

In line with this, the conductance of stomata is also highly correlated to light. This relation can be modified by environmental conditions like drought or stress by air pollution. The ratio C_i/C_a in stationary conditions is about two-thirds for C_3 but about one-third for C_4 plants. The lower value for C_4 plants reflects the higher affinity for CO_2 of the C_4. photosynthetic pathway, and reflects the more efficient water use for these plants.

Acclimation of Stomatal Movement

Little is known about the acclimation of the stomatal movement to long-term exposure to elevated [CO_2]. Some studies suggest that the lower stomatal conductance not only persists over extended exposure periods to elevated [CO_2], but is also conserved after subsequent lowering of the [CO_2] (Gorissen, pers. comm.). For glasshouse horticulture it was reported that the stomatal conductance of high-[CO_2] tomato plants was less sensitive to short-term [CO_2] fluctuations, and higher than that of plants grown at ambient concentrations and measured at double present [CO_2] (Stanghellini and Bunce, 1994). This may indicate that the sensitivity of stomatal movement of high [CO_2] plants to changes in [CO_2] is reduced, but still a lower stomatal conductivity exists when compared to leaves growing and measured in present [CO_2] conditions.

Effects at the Level of the Leaf

The partial closure of stomata is reflected in the reduced conductance at the leaf level (Atkinson *et al.*, 1991; *Sionit et al.*, 1984). At a CO_2 concentration double the present, conductance is reduced by 30-40%, although large differences among species exist.

Leaf Temperature and Vapour Pressure Deficit

Water loss by transpiration is not only affected by the conductivity of the stomata, but also by the driving forces for exchange of the water vapour from the leaf surface to the surrounding atmosphere. Therefore the gradient in partial pressure of the water vapour at the leaf surface is also of importance (McNaughton and Jarvis, 1991). All other factors being equal, the existing vapour pressure deficit (VPD) between stomatal cavity and surrounding air, the boundary layer, will increase at a reduced transpiration rate, and feed back to stimulate transpiration. The reduced transpiration will cool the leafless, and consequently a rise in temperature in the stomatal cavity and at the leaf surface may occur. Thus, in addition to the global greenhouse effect on air temperature, the temperature at the leaf surface may rise by 0.5 to 1.5 °C. In their FACE experiments (Free Air Carbon dioxide Enrichment) in Arizona, Kimball *et al.* (1995) measured an average rise in canopy temperature of 0.56°C over the growing season. Such a higher leaf temperature may also have important consequences for the longevity and photosynthetic capacity of the individual leaves and at the canopy level, as ageing may be accelerated. It is presently not known how different crops respond to the increase in leaf and canopy temperature. The mostly negative feedback on water vapour exchange will at least partly counteract the effect of the reduced stomatal conductance even without rise in ambient temperature at the global scale. A rise of ambient growing season temperatures due to the greenhouse effect will also tend to stimulate transpiration.

Depending on changes in air humidity it may affect VPD at the leaf surface determining the fluxes of sensible and latent heat and thereby the energy balance. Increased water availability may, however, offset the relative warming at the leaf surface, especially if, through preceding water saving and lower instantaneous requirements, partial stomatal closure alleviates water stress during periods of high water requirements in the diurnal sequence. It requires detailed microclimatological data and complex calculations to integrate these various mechanisms in operation at the leaf level.

Leaf Area and Weight

Changes in area of individual leaves also have a direct effect on water loss at a per leaf basis. In general a substantial increase in individual leaf weight is found. This is primarily due to an increased leaf thickness and additional accumulation of non-structural carbohydrates. Leith *et al.* (1986) found for soybeans that a small but significant increase in leaf area with CO_2 concentration may be expected (about 5% increase with a doubling [CO_2]). This is consistent with trends reported in other papers. In general, however, weight gain through increased leaf thickness is more important than that by the increase in leaf area. The relative gain in leaf area may also depend on growing conditions such as water shortage.

Effects at the Plant and Canopy Level

Aspects of the water economy at the leaf level also apply at the canopy level. Moreover, light distribution in the canopy, leaf age, humidity gradients and coupling with the atmosphere for different canopy layers and canopy structures all have their effect. The direct effects on vegetation properties discussed here are linked to the change in CO_2 concentration and temperature, but indirect effects of a climate change related with changes in ambient temperature and air humidity, coupling to the lower atmosphere and albedo changes, may even dominate. Mismanagement or over-exploitation of natural resources may lead to changes in the vegetation and even to desertification, and via a positive feedback aggravate changes in local and regional climate as studied in the EU-EFEDA-programme (European Field Experiment in Desertification-threatened Areas. This field of research is still under development and the subject of international programmes like the IGBP-BAHC (International Geosphere-Biosphere Programme - Biospheric Aspects of the Hydrological Cycle.

Leaf Area Index

The leaf area index (leaf surface area per unit soil surface area) of a crop with adequate water supply at elevated [CO_2] increases, especially early in the season, as a result of earlier and more rapid leaf production in the vegetative growth phase.

This applies especially for indeterminate growing species and under non-limiting supply of nutrients. Early development of the canopy will lead to an earlier full ground cover, and may thus limit water loss from direct soil evaporation.

Depending on the local precipitation and available soil water reserves, such an early enhanced canopy development may also be favourable for the full utilization of water resources. The higher transpiration early in the season may also lead to an earlier depletion of water reserves in the soil.

In their FACE experiments Hendrey *et al.* (1993) found that for cotton leaf area increased by 25% in dry plots, but by only 11% in irrigated plots as compared to the respective controls. Such enhancements in leaf area and biomass formation in dry conditions could improve the duration of the vegetative soil cover, and help counteract erosion and other forms of land degradation.

Plant Water Status

It has been shown that plant water status is generally improved at elevated CO_2. Part of the effect can be ascribed to a reduced transpiration demand (per unit leaf area), and therefore a partial alleviation of the water stress.

Also for plants with a normal water supply, pre-dawn and midday water potentials have been found to be less negative in high [CO_2]. Even in saline conditions the effect of partial stomatal closure is reflected in a higher salt tolerance. As a consequence, the restriction of photosynthesis under low water supply may be less severe, or be delayed.

Changes in the Rooting Pattern

Several studies have reported an increased input of carbon in below-ground processes. An important aspect to be considered in reviewing the literature at this point is that most experiments concern plant growth over a short period and with a soil volume limited by pot or container size.

In a study with container-grown grasses and clover Nijs *et al.* (1989) showed that in a rapidly developing terminal drought period water-use efficiency is about doubled, and stress is developing later under elevated [CO_2]. These conditions are thus improving the possibility to escape drought stress.

Even though the trend is clear, elevated [CO_2] plants grown with restricted soil volume may not take full advantage of a larger root system. In (semi-) field experiments the larger root system indeed leads to a better exploration of the soil volume, and an earlier, or higher, rooting density at the larger depths.

Experiments with potted plants can therefore not be expected to show the full extent to which elevated [CO_2] plants may profit in a (transient) drought period.

The observation of Morison and Gifford (1984a,b) that the pattern of depletion of soil water reserves is rather similar may partly be an underestimation of the real moisture availability in field conditions because of a better access to deeper soil layers. It has also been observed that a faster recovery of the crop is possible when grown at elevated [CO_2]. The explanation of this observation has not yet been given, but could be related to deeper and more extensive rooting patterns.

Temperature Effects

The importance of temperature at the canopy level is two-fold. Firstly higher temperatures increase transpiration by changing the VPD at the leaf surface, and secondly the higher canopy temperature may lead to an accelerated ageing of the foliage, and a shortening of the growing season, or, for example, grain-filling period. The latter type of effect is discussed extensively in crop growth modelling studies that describe changes in crop productivity due to climate change.

On theoretical grounds it can be expected that transpiration losses will increase with higher air temperatures. It is likely that the evaporative demand as determined by the vapour pressure deficit would increase by about 5 to 6% per degree warming. The overall effect of a higher temperature alone is an intensification of the hydrological cycle. It should be realized that the combined result of a higher ambient temperature, leading to a higher evaporative demand, and partial stomatal closure, counteracting this, could be overridden by changes in other environmental and atmospheric conditions like soil water availability, precipitation patterns, cloudiness and air humidity.

Effects Over the Growing Season

The effects of elevated CO_2 alone on transpiration integrated over the season show that any of the cases (higher water use; no change; lower water use) can be found. Especially in the case of non-limiting water supply a higher water use can be expected even with a lower stomatal conductance as leaf area will increase with about the same factor as total biomass.

This will compensate for the reduction in water exchange per leaf area. This applies especially in early season. In conditions with transient water shortage, profit from partial stomatal closure will be highest, as both the depletion of soil water reserves is delayed and plant sensitivity to midday water shortage (suppressed photosynthesis) is lessened.

In conditions with adequate water supply, water use over the whole season, and especially in periods in the season with a closed canopy (e.g., with leaf area index over 3), is little different if at all. ITE and WUE (24h basis) are both higher (about 40-50%) at elevated [CO_2], but transpiration per unit soil surface is not changed significantly. Here treatment temperatures have been kept unchanged, tracking outside conditions. The improvement in seasonal water-use efficiency (SWUE) ranges from 23% (wheat) to 54% (faba bean), similar to the results reported by Dijkstra *et al.*

Similarly, Baker and Allen (1993) presented results of canopy gas exchange for soybean and citrus at a range of temperatures. Although day time water-use efficiency (DWUE) fell with temperature, they measured at 800 μ l/l [CO_2] a doubling (soybean) or more of the DWUE, independent of temperature (28-35°C). Seasonal water-use efficiency varies with crop and growth conditions.

Water-use efficiency rises especially in conditions with water limitations. In their study of the potential changes of productivity of cool-season legumes, Grashoff *et al.* (1994) concluded that for rainfed faba bean crops in the Netherlands, Syria and Israel productivity would increase, especially under water limitation. Faba bean is a special case, because crop development rate is accelerated by drought.

A small increase in temperature (1.7°C) would decrease yield, but the lower water requirement at higher [CO_2] (1.7°C and 460 μ l/l for the climate in 2030) more than compensates for this, raising yields from 2.8 to 4.7 t/ha (Syria), 3.9 to 5.0 t/ha (Israel) and 5.1 to 5.7 t/ha (Netherlands).

In fully irrigated conditions yields would be higher, but the relative CO_2 effect would be lower. Interestingly the standard deviation of the predicted yield (>10 year simulated average) shows a tendency to decrease in the changed climate scenario.

Table: *Effect of [CO_2] on full canopy gas exchange parameters for spring wheat (1991) and faba bean (1992) both measured on Julian day 188.*

	Spring wheat			***Faba bean***		
[CO_2] μ mol/mol	***350***	***700***	***ratio***	***350***	***700***	***ratio***
CCERmax [1]	50.2	70.4	1.40	45.6	66.8	1.47
CDR [2]	-3.8	-6.0	1.59	-3.85	-6.50	1.69
CCER light period [7]	67.5	95.6	1.42	67.7	100.7	1.49
CDR dark period [7]	-5.4	-8.6	1.59	-5.5	-9.3	1.69
Net CCER shoot-24h [7]	62.1	87.0	1.40	62.2	91.4	1.47
CETmax [3]	9.55	8.77	0.92	7.29	7.71	1.06
CET-24h [4]	5.24	4.90	0.94	5.50	5.60	1.02
CCERmax/CETmax [5]	5.26	8.03	1.53	6.26	8.66	1.39
CCER-24h/CET-24h [6]	11.85	17.75	1.49	11.31	16.32	1.44

1. CCERmax = maximum canopy CO_2 exchange rate on a day (μ mol/m^2/s)
2. CDR = mean canopy dark (night) respiration rate (μ mol/m^2/s)
3. CETmax = maximum canopy evapotranspiration rate on a day (mmol/m^2/s)
4. CET-24h = integrated canopy evapotranspiration (mm/d)
5. Or maximum ITE in mmol CO_2/mol H20
6. Or daily WUE in g CO_2/kg H_2O
7. g CO_2/m^2/d

Table: *Daytime canopy water-use efficiency (D-WUE: photosynthesis per unit water transpired: mmol CO_2/mol H_2O)*

	[CO₂] μ mol/mol	*Temp. (°C)*	*DWUE*	*High/low ratio*
Soybean	330	28	2.95	
		31	2.63	
		33	2.37	
		35	2.33	
	800	28	6.00	2.03
		31	5.11	1.94
		33	4.95	2.09
		35	4.71	2.02
Citrus	330	25	2.30	
		34	1.25	
	840	25	5.40	2.34
		34	3.52	2.81

Table: *Seasonal crop water-use efficiency ((C)WUE in g DM/kg water)*

	Ambient [CO₂]	*Double [CO₂]*	*Ratio*	
Sorghum	3.08	4.13	1.34	
Wheat well-watered	5.1	6.3	1.23	
Wheat water shortage	6.2	8.9	1.43	
Wheat	2.62	3.45	1.31	
Wheat well-watered	1.58	2.14	1.35	grain only
Wheat water shortage	1.27	1.86	1.46	grain only
Faba beans	4.91	7.82	1.54	
Water hyacinth	1.4	2.6	1.85	

Table: *Average simulated seed yield in t/ha of rainfed and fully irrigated faba bean crops in Wageningen (Netherlands), Tel Hadya (Syria) and Migda (Israel) under present and changed climate conditions (year 2030: + 1.7°C and [CO_2] at 460 μ mol/mol; year 2080: +3.0°C and [CO_2] at 700 m mol/mol; no precipitation change).*

Location and Scenario	*Rainfed*			*Fully irrigated*		
	yield	*change (%)*	*sd (%)*	*yield*	*change (%)*	*sd (%)*
Wageningen						
Current climate	5.1	-	32	6.1	-	9
Year 2030	5.7	12	29	6.4	5	7
Year 2080	6.7	31	22	7.2	18	6
Tel Hadya						
Current climate	2.8	-	41	6.9	-	11
Year 2030	4.7	68	35	8.0	16	8
Year 2080	7.4	164	22	9.5	38	6
Migda						
Current climate	3.9	-	39	6.4	-	9
Year 2030	5.0	28	41	7.2	12	10
Year 2080	6.7	72	35	8.5	33	11

Effects on the Regional Vegetation-atmosphere Water Vapour Exchange

One of the more complex aspects of the estimation of changes in the water use of arable crops is the extrapolation from leaf and canopy to the field and region. This has long been recognized by the international scientific community, and has given rise to the establishment of several large-scale measuring campaigns (HAPEX: Hydrologic Atmospheric Pilot Experiment (Sahel and Spain), GEWEX: Global Energy and Water Cycle Experiment (World Climate Research Programme)) and to the start of the IGBP-BAHC programme. A major part of the efforts will go into the development of the SVAT-models (Soil-Vegetation-Atmosphere-Transfer models). The extrapolation from the stomatal and leaf level to the region have been the subject of experimental and theoretical studies.

In general the structure of the canopy determines to a large extent the transfer of heat and water vapour to the lower atmosphere. Most models to date neglect or oversimplify the feedback between the vegetated surface and the lower atmosphere, the planetary boundary layer (PBL).

Jacobs (1994) showed in a study using coupled models (one-dimensional model for the PBL coupled to a modified Penman-Monteith big-leaf model) that in such a model the sensitivity of regional transpiration to changes in surface resistance is reduced (about halved) but that the sensitivity to changes in albedo is increased by between 25 and 250% relative to the values obtained without PBL feedback.

In addition to the sensitivity of stomata to photosynthesis and $[CO_2]$, Jacobs (1994) introduced an additional correlative relationship between the humidity deficit at the leaf surface and the C_i/C_a ratio. The changes in transpiration caused by changes in surface resistance are damped. However, within the canopy the changes in the specific humidity lead to a positive feedback, e.g., the CO_2 effects on surface resistance are enhanced. The inclusion of PBL feedback and stomatal response to humidity deficit leads in his study to an estimated overall decrease of the regional transpiration by 10 to 30%. Modifying factors are surface roughness determined by the vegetation type, temperature and air humidity. In particular, information about the last factor, extrapolated from regionalized GCM output, is virtually absent or highly speculative.

Arable crops, especially in the temperate regions in general, have a low surface roughness, and in that case the transpiration is primarily determined by the radiation energy. This is also true for pastures. In rangelands with sparse higher vegetation the higher roughness leads to a stronger coupling, a situation comparable to forests and mixed vegetation.

Concluding Remarks

Present knowledge does not allow a firm statement to be made concerning a future situation with respect to water-limited agricultural production. At a regional production level the available quantitative data from regionalized GCM runs are insufficient to be used as direct inputs for crop growth models. At a global level the simultaneous changes in rainfall pattern, air humidity and possible shifts in vegetation zones add to uncertainty, as many and often non-linear feedbacks are expected to operate.

The limited amount of available experimental data has as a trend that water use per unit soil surface area will change little (-10 to +10%). As, however, the general trend towards an improved water-use efficiency is clear, the productivity per unit of available water is expected to rise by 20-40%, probably much less than the value (100%) calculated from a reduced stomatal conductivity and an increased photosynthesis.

Some studies show that in situations with marginal water availability the threshold for a successful crop may shift to lower values. Whether at otherwise unchanged water availability this would open up possibilities to reverse existing trends towards desertification in certain areas is very questionable, as apart from precipitation falling short to maintain existing vegetation, other aspects like over-exploitation might dominate. It should, however, be emphasized that present knowledge on feedback among vegetation characteristics, gas exchange and albedo and the regional climate is insufficient to draw firm conclusions.

Some related problems have not been touched upon. For example, changes in precipitation distribution throughout the season may lead to shifts in the accessibility of fields for farm operations in early spring or late autumn. In addition they may cause changes in soil moisture and thereby modify the rate of mineralization of nutrients (nitrogen, phosphorus). Such interactions, combined with changes in (air and soil) temperatures, may markedly change soil fertility and thereby local farming systems and crop productivity. Depending on the aim of the study, such impacts have to be considered more or less relevant to the present subject.

In many studies the impact of climate change on crop growth and yield is analysed using crop simulation models. A proper analysis of the performance of such models should be made to verify their reliability in the projected conditions of changed atmospheric composition and changed climate. The absolute values of the regionally and globally aggregated predicted crop yields, often calculated using a proportionality factor for the CO_2 response, are at present probably less reliable than the predicted sensitivity of the yield to various climate and management factors.

The above considerations relate to a projected world, changed primarily in terms of climate and atmospheric composition. It should be emphasized that at many points agricultural practice is very dynamic, and will respond to changed conditions by adaptation. Crop and cultivar choice will, in most instances and in the most productive areas, change over time and gradually incorporate the traits necessary for adapted performance, or change to better adapted species. A rising temperature will for most current cultivars, for instance, accelerate crop development. This would in itself lead to a reduced water use over the shortened growth period, but also to a loss of potential yield. One may argue that farmers will repair such a loss of production potential by a proper choice of adapted cultivars or crop species, unless temperatures exceed the appropriate temperature window.

These developments require, however, that new technologies and genetic resources be practically and economically accessible for all farmers, a situation that is not reached at present for farmers in arid and semi-arid regions in developing countries where the risks but probably also opportunities for agricultural production may be greatest.

Effects of Higher Day and Night Temperatures on Growth and Yields of Some Crop Plants

Gaseous emissions from human activities are substantially increasing the concentrations of atmospheric greenhouse gases, particularly carbon dioxide, methane, chlorofluorocarbons and nitrous oxides. Global circulation models predict that these increased concentrations of greenhouse gases will increase average world temperature. Under the business-as-usual scenario of the Intergovernmental Panel on Climate Change (IPCC), global mean temperatures will rise 0.3°C per decade during the next century with an uncertainty of 0.2 to 0.5%. Thus global mean temperatures should be 1°C above the present values by 2025 and 3°C above the present value by 2100. Although global circulation models do not all agree as to the magnitude, most predict greenhouse warming. There is also general agreement that global warming will be greater at higher latitudes than in the tropics. Different global circulation models have predicted that global warming effects will vary diurnally, seasonally and with altitude.

It is also possible that there will be an autocatalytic component to global warming. Photosynthesis and respiration of plants and microbes increase with temperature, especially in temperate latitudes. As respiration increases more with increased temperature than does photosynthesis, global warming is likely to increase the flux of carbon dioxide to the atmosphere which would constitute a positive feedback to global warming.

This chapter describes the effects of higher day and night temperatures on crop growth and yield. Temperature effects at different levels of organization - biochemical, physiological, morphological, agronomic and systems - are considered. This is followed by identification of options for germplasm improvement and crop management that may mitigate the adverse effects of higher day and night temperatures. The main focus is on wheat (*Triticum aestivum* L.) and rice (*Oryza saliva* L.).

Mechanisms for Heat Tolerance

Crop plants are immobile. They must adapt to prevalent soil and weather conditions. Except for transpirational cooling, plants are unable to adjust their tissue temperatures to any significant extent. On the other hand, plants have evolved several mechanisms that enable them to tolerate higher temperatures. These adaptive thermotolerant mechanisms reflect the environment in which a species has evolved and they largely dictate the environment where a crop may be grown.

Four major aspects of thermotolerance have been studied: (1) thermal dependence at the biochemical and metabolic levels; (2) thermal tolerance in relation to membrane stability; (3) induced thermotolerance through gradual temperature increase vis-a-vis production of heat shock proteins; and (4) photosynthesis and productivity during high temperature stress.

Biochemical Processes

Temperature effects on the rates of biochemical reactions may be modelled as the product of two functions, an exponentially increasing rate of the forward reaction and an exponential decay resulting from enzyme denaturation as temperatures increase. The greatest concern is whether it is possible to increase the upper limit of enzyme stability to prevent denaturation.

Failure of only one critical enzyme system can cause death of an organism. This fact may explain why most crop species survive sustained high temperatures up to a relatively narrow range, 40 to 45°C. The relationship between the thermal environment for an organism and the thermal dependence of enzymes has been well established.

The shape of this function also describes temperature effects on most biological functions, including plant growth and development. The function can be categorized by the three cardinal temperatures - minimum, optimum and maximum. Modellers frequently simplify the relationship into a stepwise linear function. The stepwise linear function has a plateau rather than an optimum temperature.

The thermal dependence of the apparent reaction rate for selected enzymes may indicate the optimal thermal range for a plant. The range over which the apparent Michaelis-Menten constant for CO_2 (K_m) is minimal and stable is termed the thermal kinetic window. For crop plants, the thermal kinetic window (TKW) is generally established as a result of thermally induced lipid phase changes, rubisco activity and the starch synthesis pathway in leaves and reproductive organs.

In cotton and wheat, the time during which foliage temperature remained within the TKW was related to dry matter accumulation. The cumulative time that rainfed crop foliage is outside the TKW provides an index of the degree of extreme temperature stress of the environment. Irrigation is one management option to reduce crop exposure to heat stress.

Temperatures that inhibit cellular metabolism and growth for a cool season C_3 species such as wheat may not inhibit warm-season C_3 species such as rice (*Oryza sativa* L.) and C_4 species such as sorghum, maize (*Zea mays* L.) and sugar cane (*Saccharum spontaneum* spp.). The identification of TKWs for different species can aid in the interpretation of the differential temperature stress responses for crop growth and development among species.

Thermal Stability of Cell Membranes

The plasmalemma and membranes of cell organelles play a vital role in the functioning of cells. Any adverse effect of temperature stress on the membranes leads to disruption of cellular activity or death. Heat injury to the plasmalemma may be measured by ion leakage. Injury to membranes from a sudden heat stress event may result from either denaturation of the membrane proteins or from melting of membrane lipids which leads to membrane rupture and loss of cellular contents.

Heat stress may be an oxidative stress. Peroxidation of membrane lipids has been observed at high temperatures, which is a symptom of cellular injury. Enhanced synthesis of an anti-oxidant by plant tissues may increase cell tolerance to heat but no such anti-oxidant has been positively identified.

A relationship between lipid composition and incubation temperature has been shown for algae, fungi and higher plants. In *Arabiodopsis,* exposed to high temperatures, total lipid content decreases to about one-half and the ratio of unsaturated to saturated fatty acids decreases to one-third of the levels at temperatures within the TKW. Increase in saturated fatty acids of membranes increases their melting temperature and thus confers heat tolerance. An *Arabiodopsis* mutant, deficient in activity of chloroplast fatty acid W-9 desaturase, accumulates large amounts of 16:0 fatty acids, resulting in greater saturation of chloroplast lipids. This increases the optimum growth temperature.

In cotton, however, heat tolerance does not correlate with degree of lipid saturation (Rikin *et al.,* 1993) and similar differences in genotypic differences in heat tolerance have been unrelated to membrane lipid saturation in other species. In such species, a factor other than membrane stability may be limiting growth at high temperature.

Heat Shock Proteins

Synthesis and accumulation of proteins were ascertained during a rapid heat stress. These were designated as 'Heat Shock Proteins'

(HSPs). Subsequently it was shown that increased production of these proteins also occurs when plants experience a gradual increase in temperature more typical of that experienced in a natural environment.

Three classes of proteins as distinguished by molecular weight account for most HSPs, namely HSP90, HSP70, and low molecular weight proteins of 15 to 30 kDa (LMW HSP). The proportions of the three classes differ among species. In general, heat shock proteins are induced by heat stress at any stage of development. Under maximum heat stress conditions, HSP70 and HSP90 mRNAs can increase ten-fold and LMW HSP increase as much as 200-fold. Three other proteins, though less important, are also considered to be heat shock proteins viz. 110 kDa polypeptides, ubiquitin, and GroEL proteins.

In arid and semi-arid regions, dryland crops may synthesize and accumulate substantial levels of heat shock proteins in response to elevated leaf temperatures. The induction temperature for synthesis and accumulation of heat shock proteins in laboratory-grown cotton ranged from 38 to 41°C. Soil water deficits resulting in midday canopy temperature of 40°C or greater for two to three weeks were used to study heat shock proteins in field-grown cotton. A comparison of polypeptide patterns of dryland and irrigated cotton leaves showed that at least eight new polypeptides accumulated in about half of the dryland leaves analysed. The polypeptides that accumulated in the dryland leaves but not irrigated cotton leaves had molecular weights of 100, 94, 89, 75, 60, 58 and 21 kDa. In a similar experiment with field-grown soybean (*Glycine max* (L.) Merr.), several heat shock proteins were observed in both irrigated and dryland treatments, although levels were greater in the non-irrigated treatments.

Correlation between synthesis and accumulation of heat shock proteins and heat tolerance suggests, but does not prove, that the two are causally related. Further evidence for a causal relationship is that some cultivar differences in heat shock protein expression correlate with differences in thermotolerance. In genetic experiments, heat shock protein expression co-segregates with heat tolerance. Another evidence for the protective role of heat shock protein is that mutants unable to synthesize heat shock proteins, and cells in which HSP70 synthesis is blocked or inactivated, are more susceptible to heat injury.

The mechanism by which heat shock proteins contribute to heat tolerance is still not certain. One hypothesis is that HSP70 participates in ATP-dependent protein unfolding or assembly/disassembly reactions and that they prevent protein denaturation during stress. If this mechanism is true, then heat shock proteins may provide a significant basis for increasing heat tolerance of crop plants in a global warming situation. The LMW HSPs may play a structural role in maintaining cell membrane integrity during stress.

Other heat shock proteins have been associated with particular organelles such as chloroplasts, ribosomes and mitochondria. In tomato (*Lycopersicon esculentum* L.), heat shock proteins aggregate into a granular structure in the cytoplasm, possibly protecting the machinery of protein synthesis.

HSPs provide a significant opportunity to increase heat tolerance of crops. To elucidate their mechanisms of action and to exploit their potential contribution to increasing heat tolerance, four lines of investigations are suggested:

1. Establish the biochemical activities of individual HSPs as a preliminary step.
2. Characterize the genetic variability of specific heat shock proteins across a wide range of germplasm. Develop iso-population and near isogenic lines selected for production of low and high levels of HSP synthesis.
3. As HSPs appear to participate in maintaining the conformation or assembly of other protein structures, analyse the molecular details of these processes and establish all participating protein substrates. Such biochemical studies are needed to understand how these processes protect or allow recovery from heat stress.
4. Identify specific HSP mutants or create transgenic mutant plants to complement molecular and biochemical understanding with genetic approaches.

Photosynthesis and High Temperature Stress

Variability in leaf photosynthetic rates within or between species is often unrelated to differences in productivity. Similarly, high photosynthetic rates at high temperatures do not necessarily support high rates of crop dry matter accumulation. The temperature optimum for photosynthesis is broad, presumably because crop plants have adapted to a relatively wide range of thermal environments. A 1 to 2°C increase in average temperature is not likely to have a substantial impact on leaf photosynthetic rates. Further, there is a possibility that photosynthesis of crop plants can adapt to a slow increase in global average temperatures. Thus, global warming is not likely to affect photosynthetic rates per unit leaf area gradually or on a closed canopy basis over the next century.

While photosynthetic rates were found to be temperature-sensitive in other crops, wheat and rice appear to be different. In wheat, no measurable differences were found in photosynthetic rates per unit flag leaf area or on a whole-plant basis in the temperature range from 15 to 35°C. In rice, there is little temperature effect on leaf carbon dioxide assimilation from 20 to 40°C.

Recent research has shown significant variation among wheat cultivars with respect to reduction in photosynthesis at very high temperature. Photosynthesis of germplasm adapted to higher temperature environments was less sensitive to high temperature than was germplasm from cooler environments. When this germplasm was grown under moderate (22/17°C) and high (32/27°C) temperatures in the seedling stage or from anthesis to maturity, there was a highly significant correlation between photosynthesis rate and either seedling biomass (r=0.943***) or grain yield of mature plants (r=0.807**). Genotypes most tolerant to high temperatures had the most stable leaf photosynthetic rates across temperature regimes or they had the longest duration of leaf photosynthetic activity after anthesis and high grain weights. The above relationship was exemplified by 'Ventnor' from the high temperature area of Australia and 'Lancero' from the high altitude area of Chile. Despite observed negative effects of high temperature on leaf photosynthesis, the temperature optimum for net photosynthesis is likely to increase with elevated levels of atmospheric carbon dioxide. Several studies have concluded that CO_2-induced increases in crop yields are much more probable in warm than in cool environments. Thus, global warming may not greatly affect overall net photosynthesis.

Night Respiration Response To Co_2 and Temperature

In tomato (*Lycopersicon esculentum*) and cotton (*Gossypium hirsutum*), elevated CO_2 increased dark respiration possibly because of increased carbohydrate accumulation in tissues. The latter has been shown to increase alternative pathway respiration as well.

Apparent dark respiration may decline under elevated CO_2 if there is dark CO_2 fixation or if elevated CO_2 directly inhibits or inactivates respiratory enzymes as may occur through increased formation or carbamate.

Table: *Mean weekly photosynthetic rate (fmol* $CO_2/m^2/s$*) and duration of photosynthetic activity (weeks, in parentheses), and grain biomass of two wheat genotypes grown at two temperature regimes*

GENOTYPE	*Seedlings after two weeks of treatment*		*From anthesis to maturity*		*Grain biomass (g/tiller)*	
	22/17 °C	*32/27 °C*	*22/17 °C*	*32/27°C*	*22/17 °C*	*32/27 °C*
Ventor	8.6	7.0	6.4(7)	5.3 (7)	0.93	0.8
Lancero	7.9	4.6	4.1 (10)	3.1(7)	0.43	0.28

Figure: *Figures in parentheses give duration of photosynthetic activity in weeks from anthesis to physiological maturity.*

Few studies have successfully partitioned the effects of elevated CO_2 on growth and maintenance respiration. Both components appear to decline, probably because of decrease in leaf protein levels which results in reduced construction and maintenance costs.

Elevated CO_2 reduced maintenance respiration of *Medicago sativa* and *Dactylis glomerata* at lower temperatures (15 to 20°C), whereas elevated CO_2 reduced growth respiration of *M. sativa* at 20 to 30°C and *D. glomerata* at 15 to 25°C.

Crop Growth and Development

Wheat development is customarily divided into vegetative and reproductive phases, with either ear emergence or anthesis as the event that separates the two phases. In the past 30 to 40 years, the sequence of pre-anthesis phenological events has been critically assessed with respect to grain yield potential and sensitivity to weather variables, particularly prevailing temperature and day length. Several systems are used to classify the sequence of phenological events. We identify five such developmental stages:

(i) germination - seeding to seedling emergence;

(ii) canopy development - emergence to first spikelet initiation, the double ridge stage;

(iii) spikelet production - first spikelet initiation to terminal spikelet formation;

(iv) spikelet development - terminal spikelet formation to anthesis;

(v) grain development - anthesis to maturity.

These stages are generally based on early recognized features of the apical meristem. They mark significant changes in morphology or physiology of different crop organs. Numbers of leaf and tiller primordia are determined before spikelet initiation but their subsequent growth and development are controlled by temperature and day length during the differentiation of spikes into spikelets. Similarly, floret number within each spikelet is established by anthesis, at which time the potential grain number per spike is established.

Productivity of wheat and other crop species falls markedly at high temperatures. Wheat in India is invariably exposed to extreme temperatures during some stages of development (Abrol *et al.*, 1991). In Australia, wheat is usually exposed to brief periods of heat stress during grain development. All stages of development are sensitive to temperature. It is the main factor controlling the rate of crop development. Development generally accelerates as temperature increases, a phenomenon that is often described as a linear function of daily average temperature. The growing degree day concept is a common example of a linear model of developmental response to temperature. While a linear model works well to describe wheat development as long as temperatures remain within 10 to 30°C, a non-linear model as is needed to describe development when a crop is exposed to extreme temperature stress.

Vegetative Phase

Several experiments have observed the effects of temperature on the duration from sowing or emergence to heading under controlled environment and field conditions. Unfortunately, however, few experiments have been conducted with enough cultivars to assess the genetic variability in this trait. The major conclusions from these studies are:

1. All genotypes are sensitive to temperature at one stage or another. Temperature sensitivity, however, varies greatly with genotype.
2. Phenological stages differ in sensitivity to temperature.
3. The duration of phase from sowing to first spikelet initiation is less sensitive to change in temperature than are other phases, although genotypes do differ in thermotolerance during this phase.
4. The stages during which environment has the greatest impact on yield are from first spikelet initiation or terminal spikelet formation until anthesis. Spikelet number and floral number (potential grain number), both dominant yield contributing attributes, are established during these phases. Grain weight,

on the other hand, appears to be much less sensitive to heat stress than is grain number.

Table: *Response of phasic development to temperature photoperiod and vernalization*

Developmental phase	*Temperature*	*Photoperiod*	*Vernalization*
Sowing-emergence	++++/+++++	0	0
Emergence-double ridges	+++/+++++	0/+++	0/+++++
Double ridges-terminal spikelet	+/+++	0/+++++	0/+++++
Terminal spikelet-heading	+++/+++++	0/+++	0/+(?)
Heading-anthesis	+++/+++++	0/+(?)	0
Anthesis-maturity	+++/+++++	0	0

For the estimation of sensitivity, the total life span was divided into the stages. An arbitrary scale was used to show when the effects are strong (+++++), moderate (+++) or slight (+). 0 denotes that the factor does not affect the process and question marks refer to uncertainties in the literature. For each factor, genetic variation in response was considered.

Grain Development Phase

In experiments under controlled conditions from 25 to 35°C, mean grain weight declined 16% for each 5°C increase in temperature. In pot experiments, grain yield decreased by 17% for each 5°C rise. For every 1°C rise in temperature, there is a depression in grain yield by 8 to 10%, mediated through 5 to 6% fewer grains and 3 to 4% smaller grain weight.

To elucidate the causal factor for reduced grain filling in wheat because of higher temperatures, Wardlaw (1974) studied the three main components of the plant system. The three components are: (a) source - flag leaf blade; (b) sink - ear; and (c) transport pathway - peduncle. He observed that photosynthesis had a broad temperature optimum from 20 to 30°C with photosynthesis declining rapidly at temperatures above 30°C. The rate of 14C assimilate movement out of the flag leaf, phloem loading, was optimum around 30°C; the rate of 14C assimilate movement through the stem was independent of temperature from 1 to 50°C. Thus, in wheat, temperature effects on translocation result indirectly from direct temperature effects on source and sink activities.

In a subsequent experiment with source-sink relationships altered through grain excision, defoliation and shading treatments, heat stress still reduced grain weight. This result supports the earlier findings that temperature effects on grain weight are direct effects rather than assimilate availability. Furthermore, respiration effects do not appear to be the direct cause of decreased grain size in heat-stressed wheat.

Reduction of grain weight by heat stress may be explained mostly by effects of temperature on rate and duration of grain growth. As temperature increased from 15/10°C to 21/16°C, duration of grain filling was reduced from 60 to 36 days and grain growth rate increased from 0.73 to 1.49 mg/grain/day with a result of minimal influence on grain weight at maturity.

Further increase in temperature from 21/16°C to 30/25°C resulted in decline in grain filling during 36 to 22 days with a minimal increase in grain growth rate from 1.49 to 1.51 mg/grain/day. Thus, mature grain weight was significantly reduced at the highest temperature.

Research on the effects of brief periods of ear warming after anthesis on ear metabolism have identified differential responses of starch and nitrogen accumulation in grain of four wheat cultivars. Warming increased the rate of dry matter accumulation in all the cultivars but the increase was less in cv. Aus 22645 than in the other cultivars studied.

Rate of increase in nitrogen accumulation was, however, higher than the increase in total dry matter accumulation. Under long-term exposure to heat stress, increased grain nitrogen concentration is almost entirely as a result of decreased starch content rather than a change in total grain quality. The conversion of sucrose to starch within the endosperm is decreased by elevated temperatures. Furthermore, heat stress effects on final grain weight were associated with reduced levels of soluble starch synthetase activity.

In summary, high temperature reduction of grain yield results from: (a) reduced numbers of grains formed; (b) shorter grain growth duration; and (c) inhibition of sucrose assimilation in grains.

Extreme Temperature Effects on Crops

There are two major forms of extreme temperature stress on crops - heat and cold. An increase in global temperatures may have either or both of these two acute effects: more frequent high temperature stress and less frequent cold temperature stress.

Increase in temperature will lengthen the effective growing season in areas where agricultural potential is currently limited by cold temperature stress. Thus, increased temperature will cause a poleward shift of the thermal limits to agriculture. This poleward shift will be especially important for crops such as rice that have tropical centres of origin and adaptation but are also grown in temperate latitudes during warm seasons. Global warming impact will be greater in the northern than southern hemisphere because there is more high-latitude area cultivated in the northern hemisphere.

Table: *Effect of whole-plant warming on the rate of total dry matter and nitrogen accumulation, between days 10 and 20 after anthesis, in the grains of four cultivars of wheat*

Cultivar	*Treatment*	*Rate of increase (mg/grain/day)*	
		Total dry matter	*N content*
AUS 22645	C	1.94±09	0.03±002
	W	2.07±07(107)	0.48+003(160)
Kite	C	1.72+15	0.027±004
	W	2.28±17(133)	0.043±003(159)
Sonora	C	1.65±18	0.034±009
	W	2.06±19(125)	0.051±009(150)
WW15	C	1.89±20	0.037±005
	W	2.37±20(125)	0.053+005(143)

Plants were grown at 21/16°C and some (W) were warmed, between 10 and 20 days after anthesis, to 33/25°C and then returned to 21/16°C where they stayed until maturity. Control plants (C) were grown at 21/16°C throughout. Values given are the means ±.e; values in parentheses are percentage of cotton values.

Increased temperature would also affect the crop calendar in tropical regions. In the tropics, however, global warming, though predicted to be of only small magnitude, is likely to reduce the length of the effective growing season, particularly where more than one crop per year is grown. In semi-arid regions and other agro-ecological zones where there is wide diurnal temperature variation, relatively small changes in mean annual temperatures could markedly increase the frequency of highest temperature injury. For example, canopy temperature is 10 to 15°C higher in dryland cotton (*Gossypium hirsutum* L.) than in irrigated cotton. Thus, global warming would reduce dry matter accumulation in dryland cotton because of increased respiration, and reduced photosynthesis and cellular energy.

In India, the growing season for wheat is limited by high temperatures at sowing and during maturation. As wheat is grown over a wide range of latitudes, it is frequently exposed to temperatures above the threshold for heat stress. For example, rainfed wheat depends on soil moisture remaining after the monsoon rains recede in September. High maximum and minimum temperatures in September (about 34/20°C), which adversely affect seedling establishment, accelerate early vegetative development, reduce canopy cover, tillering, spike size and yield.

Hence, sowing is typically delayed until after mid-October when seedbeds have cooled, though much of the residual soil moisture may be lost. High temperatures in the second half of February (25/10°C), March (30/13°C) and April (30/20°C) reduce the numbers of viable florets and the grain-filling duration. High temperature stress particularly reduces yield of wheat sown in December/January which is necessitated in some regions because of the multiple cropping system.

The situation is similar for sorghum (*Sorghum bicolour* (L.) Moench) and pearl millet (*Pennisetum glaucum* (L.) R.Br.) which are exposed to extreme high temperatures in Rajasthan, India. After sowing, air and soil temperatures often exceed 40°C and midday soil surface temperatures above 50°C are common.

Acute effects of high temperature are most striking when heat stress occurs during anthesis. In rice, heat stress at anthesis prevents anther dehiscence and pollen shed, to reduce pollination and grain numbers. Clearly, many crops in tropical areas are already subjected to heat stress. If temperatures increase further, crop failure in some traditional areas would become more commonplace.

Long-term Effects of High Temperatures on Crops

More important than acute effects of extreme temperature stress are the chronic effects of continuously warmer temperatures on crop growth and development. Chronic effects of high temperature include effects on grain growth discussed above. Record crop yields clearly reflect the importance of season-long effects on crop yields: crops generally yield the most where temperatures are cool during growth of the harvested component.

Crop growth simulations show that rice yields decrease 9% for each 1°C increase in seasonal average temperature. This chronic effect of high temperature differs significantly from the acute effect of short-term temperature events, because seasonal temperature effects are mostly a result of effects on crop development. For most grain crops, there is much greater genotypic variation in thermal requirements for vegetative than for reproductive development. As long-term temperatures increase, grain-filling periods decrease, and there appears to be little scope to manipulate this effect through existing genetic variation within species.

Figure: *Diurnal temperature data recorded in Fatehpur, Rajasthan, India. (Latitude 27°C 37'N) in June 1989). Each measurement is the mean value from three thermocouples placed at either 5 cm depth of soil (▲): 0.5 cm depth of soil (●): or 150 cm above the soil surface (■).*

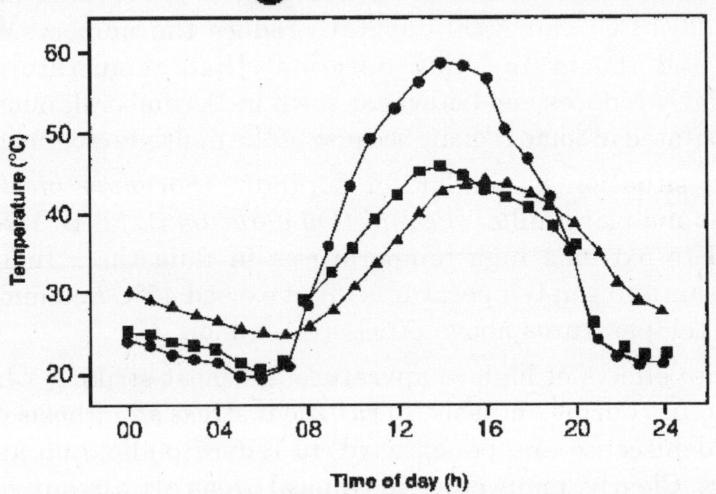

Combined Effects of Changing CO_2 Temperature, UV-B radiation and O_3 on Crop Growth

Our understanding of the relationships between crop growth and the atmospheric environment has developed substantially in the past few decades. Improvements in technology have enabled crop physiologists to study processes such as transpiration and carbon dioxide exchange in the field. Mechanistic understanding of the factors controlling photosynthesis and respiration has improved greatly, but we still have no mechanistic understanding of the ways in which many abiotic stresses affect these processes. Our knowledge of factors that control crop development and the partitioning of assimilate between organs is still at the descriptive stage, and poses one of the main limitations on our ability to develop fully mechanistic mathematical models of crop growth. Nevertheless, crop science has reached a stage where reliable semi-empirical simulation models of some of the major world crops have been constructed and used to explore the potential for crop growth and yield in different environments. For the reasons discussed above, these models are restricted in their application to situations where the empirical relationships in them are well defined, but they are often useful in the design of particular experiments to investigate new environments, for example, to apply them to scenarios of climatic change. It should be emphasized that, although we have a much improved understanding of the growth and yield potential of major agricultural crops, this knowledge needs to be extended to include other crops, especially those important for developing countries.

However, as we steadily improve our understanding of how crop yield is influenced by the atmospheric environment, that environment is constantly changing as a result of human activities, and the challenge today is to predict how crops will respond to the changed environments of tomorrow. Traditionally, crop scientists have used controlled environments and field-based facilities to investigate crop responses to individual factors, such as radiation, temperature, humidity and water availability.

More recently, manipulation of air quality has been used to study crop responses to air pollutants, such as SO_2, O_3 NO_x, and some of the technology developed for these experiments has been successfully applied to study responses to increased CO_2. Although this 'single component' approach is a highly effective way of discovering how one atmospheric factor influences crops, in practice the atmosphere in the future will alter in several ways simultaneously.

Some of the impending changes are more certain than others. Without doubt, CO_2 concentration is increasing worldwide at about 1.8 μ mol/mol per year and under all likely scenarios will continue to do so. This increase will enhance the global greenhouse effect, and the consensus of scientific opinion is that it will result in global warming and in changed global distributions of rainfall and other weather components.

The amount of climatic change is, however, very uncertain, because even the best computer simulation models do not adequately describe features such as ocean-atmosphere linkages, and feedbacks associated with changes in atmospheric aerosols or cloud amount and distribution. At present, expert opinion is that, when the CO_2 concentration reaches double its present value of about 350 μ mol/mol, global temperatures will be about 2.5°C warmer, but this figure and the associated global variability are very uncertain; even more uncertain are any changes in rainfall amount and distribution.

In contrast to the uncertainty associated with computer models, there is clear evidence from satellite and from surface monitoring that stratospheric ozone is being depleted, probably because of emissions of chlorine- and bromine-containing substances, so that globally the column of ozone from the surface through the atmosphere is thinning at about 3% per decade, a rate that seems to be increasing.

Because the chemicals believed to be causing this thinning are long-lived in the stratosphere, this thinning process is likely to continue for several decades even though the chemicals are being phased out of production. The consequence of this thinning is that ultraviolet (UV) radiation at the ground, normally severely attenuated by stratospheric O_3 will increase.

There is, however, an inadequate network of surface monitoring stations to define the UV climate at the ground (i.e., taking into consideration effects of cloud, dust, etc.) or to determine whether UV has increased significantly. Even if annual mean UV exposure has not changed, there are indications that thinning of stratospheric ozone, similar but less extreme than that observed in the Antarctic, can occur episodically at temperate latitudes.

Air quality in the troposphere is also changing. In particular, in many highly populated regions, emissions of hydrocarbons and exhaust gases from motor vehicles provide the precursors for photochemical production of tropospheric ozone. There is evidence that background tropospheric ozone concentrations have increased by about 10% per decade in Europe over the past 20 years, and that the European background has approximately doubled since the beginning of the century. Although there are no equivalent long-term records, it is likely that the tropospheric O_3, concentration in the United States and Southeast Asia has also increased. Perhaps more important than the mean background concentration of O_3 is the regular occurrence on a regional scale of high concentrations of tropospheric O_3 (typically 5 to 10 times background) associated with stagnant weather systems in summer. These concentrations have been demonstrated to reduce the growth and yield of many agricultural crops.

The atmospheric environment of many world crops of the future will undoubtedly contain more CO_2 than present, is likely to be somewhat warmer, and rainfall (and hence humidity) and cloudiness will probably also change, though in a direction unknown at present. Whether UV will increase much on a seasonal basis remains unclear, but episodes of increased UV at the ground, associated with patches of thinned stratospheric O_3, may occur, especially in temperate latitudes. Without doubt, in regions with large population densities, episodes of high concentrations of tropospheric O_3 and of other pollutants associated with fossil-fuel burning will become more common.

In this chapter we review our knowledge of some of the interacting effects of these changing atmospheric conditions on agricultural crops, and we draw attention to particular gaps where further research is needed. In an attempt to structure what could otherwise be a very disparate summary, we will concentrate on two major gases, CO_2 and O_3, and will describe work that has investigated the effects of other stresses combined with them. In each case, we first briefly review responses to these gases alone.

Carbon Dioxide

Several recent reviews of responses of crops to CO_2 have been published, and there is no advantage in repeating the detail of these here. At the leaf level, the two most well-known responses to elevated CO_2 are an increase in the rate of net photosynthesis, P_N, and a decrease in stomatal conductance, g_s.

The increase in P is much larger in C_3 species (50 to 100% when CO_2 concentration doubles) than in C_4 species (10%), but a substantial decrease in g (30 to 40%) is observed in both types of species. There is some evidence of photosynthotic acclimation to elevated CO_2, so that increased rates of P_N do not persist at the leaf level in long-term studies with some species, but there is no evidence of acclimation in g. It is not clear whether increased CO_2 also indirectly affects plant growth, e.g., by changing rates of leaf initiation, expansion or longevity, but such effects could have a large influence on crop productivity.

Much of our understanding of effects of CO_2 on plants has been gained from studies with individual leaves. There have been far fewer studies of long-term effects of elevated CO_2 on canopy-scale photosynthesis and transpiration. Most of those reported have been in closed- or open-top chambers, though a few reports of field-scale exposures (FACE) are now becoming available. Drake and Leadley summarized the data available at that time and concluded that: (1) canopy photosynthesis increases in elevated CO_2 when there is a sink available for the carbon; (2) the relative effect of CO_2 is greatest at highest temperatures; and (3) elevated CO_2 alters many interacting factors, such as canopy architecture and partitioning of assimilates, that mediate gas exchange of canopies and ecosystems.

As a consequence of increased net photosynthesis, dry matter production and yield are substantially increased by elevated CO_2. Several authors (e.g., Lawlor and Mitchell, 1991) have reviewed the literature and concluded that, provided there is adequate water, nutrients and pest control, yields of C_3 and C_4 crops growing in about 700 μ mol/mol CO_2 would be about 30 to 40% and 9%, respectively, greater than present yields (350 μ mol/mol CO_2) if all other climatic factors were unchanged.

However, in much of the world, ideal crop growing conditions, with adequate water and nutrition, are wishful thinking, and the influence of elevated CO_2 on aspects such as water-use may be much more relevant. Reviews by Morison (1985) and Eamus (1991) make it clear that, on a leaf area basis, elevated CO_2 improves the water-use efficiency (WUE - the ratio of weight of dry matter produced to weight of water transpired) of plants. On a whole-plant basis, the increase in leaf area resulting from enhanced photosynthesis may cause the amount of water-used per plant to remain close to that of plants growing at current levels of CO_2, although the WUE as defined here increases. We return to this topic later.

CO_2 and Temperature

As discussed earlier, the atmospheric environment of the future is likely to include correlated increases of atmospheric CO_2 concentration and temperature. Long (1991) reviewed the mechanisms by which temperature and CO_2 affect photosynthesis in C_3 species, and developed models of the response of leaf canopy carbon exchange to changes in these variables. He showed that the interaction of carbon dioxide concentration and temperature causes the temperature optimum of the light-saturated rate of CO_2 uptake to increase as CO_2 increases. Without this interaction, an increase in global temperature of 2.5°C would increase the frequency of temperatures in temperate regions that are supra-optimal for light-saturated photosynthesis. However, if warming on this scale was coincident with a CO_2

concentration increase to above 500 µ mol/mol, the temperature optimum would increase, and this inhibition would not occur. Long went on to develop a simple model of canopy CO_2 uptake (Ac) in response to temperature, taking into account the substantial fraction of the canopy leaf area that would be shaded and therefore not photosynthesizing at the saturated rate. Using a simple model of the diurnal variation of radiation and temperature at different latitudes, Long showed that the implications of daily net carbon assimilation responds much more to increased CO_2 at low latitudes than at high latitudes.

Table: *Harvest indices (HI, seed or pod yield per plant/total dry mass per plant) and seed or pod yields for stands of groundnut* (Arachis hypogaea *L.) and sorghum* (Sorghum bicolour *L.) grown in controlled-environment glasshouses at 350 m mol/mol or 700 m mol/mol CO_2*

Crop	*HI*		*Pod/seed yield (g/plant)*		*Percentage yield increase at elevated CO_2*
	350 µ mol/mol CO_2	*700 µ mol/mol CO_2*	*350 µ mol/mol CO_2*	*700 µ mol/mol CO_2*	
Groundnut	0.195	0.212	9.06	13.29	30.9
	±0.019	±0.038	±0.06	±1.70	±13.3
Sorghum	0.333	0.334	25.78	27.37	6.2
	±0.016	±0.015	±0.11	±0.85	±3.3

Idso grew crops in field chambers in Phoenix, Arizona, at 300 and 600 µ mol/mol CO_2. Figure shows the ratio of net photosynthesis at the two CO_2 concentrations plotted against leaf temperature, and includes values calculated from Long's theoretical relationships, showing the good agreement at the leaf level. For a number of crops grown in open-top chambers in Phoenix, Arizona, summarizes the relative increase in growth (weekly dry matter production, comparing 650 µ mol/mol CO_2 with 350 µ mol/mol) with mean air temperature, indicating that, at low temperatures, elevated CO_2 actually decreased growth (a phenomenon predicted by Long's models).

Interactions between temperature and CO_2 also need to account for effect of temperature on development. Squire and Unsworth (summarized in Goudriaan and Unsworth, 1990) pointed out that determinate crops with discrete elements to their life cycle develop faster in higher temperatures, and so the stage of seed filling is shortened, limiting the benefits of elevated *CO_2*. When CO_2 concentration doubled, but daily weather data from a typical year were used, the potential grain yield was 27% larger than in a control run at 340 µ mol/mol CO_2 (11.5 and 9.0 t/ha, respectively). The date of maturity was unchanged. When daily temperatures were increased by 3°C and CO_2 was doubled, more rapid development of the crop shortened the growing season, and the potential grain yield (10.4 t/ha) was only 15% larger than in the control, but maturity occurred 30 days earlier. The wheat model used in these simulations contains a number of elements that are sensitive to temperature. Recently, Kocabas completed a detailed investigation of the model sensitivity to changes in mean temperature at various stages of crop development, and to changes in variability of temperature in various developmental phases and over the whole season. The results indicated the particularly strong dependence of yield on temperature in the phase from emergence to double ridges in wheat, but suggested that changes in temperature variability of ±20% would not have any consistent effects on yield variability. It would be useful to incorporate Long's analysis into the wheat model to allow more precise simulation of the interactive effects of temperature and CO_2 on carbon assimilation and crop development.

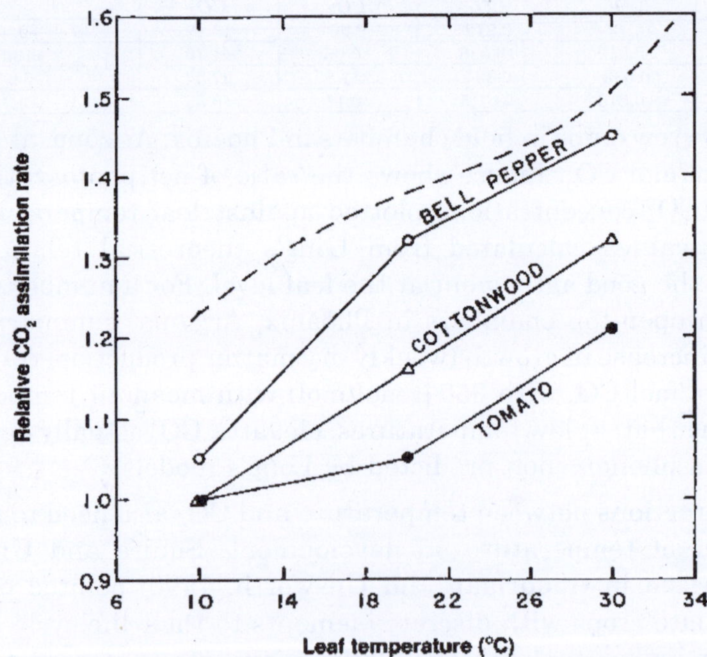

Figure: *Variation of relative CO_2 assimilation rate (net photosynthesis at an intercellular CO_2 concentration of 600 μ mol/mol divided by net photosynthesis at 300 μ mol/mol) with leaf temperature for bell pepper, tomato and cottonwood tree. The dotted line is the relationship predicted by the analysis of Long*

Indeterminate crops, such as grass and sugar beet, continue to grow and produce yield as long as the temperature is above a minimum threshold, provided that other factors such as drought and nutrition do not limit growth. Consequently, simultaneous increases in temperature and CO_2 concentration increase the potential yield of such crops in temperate environments, where low temperatures at the beginning and end of the season usually limit productivity.

Perennial tree crops, which yield fruit, nuts and wood, pose complex problems in assessing the effects of CO_2 and temperature interactions, because their yield depends on a phased sequence of development over at least two years. Climatic warming could disrupt this sequence, for example, by providing insufficient winter chilling to synchronize spring bud break, or by advancing ovule development and leading to poor fruit set in spring. Cannell and Smith (1986) pointed out that climatic warming could either advance or delay bud burst in apples depending on the extent to which chilling requirements were met. The long-term influence of CO_2 on tree growth, and its interaction with temperature responses, is very uncertain at present.

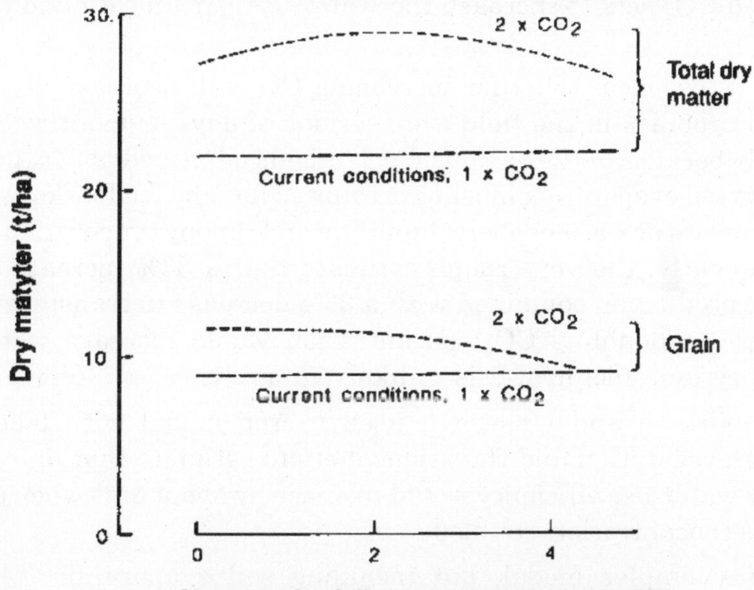

Figure: *Calculations using the model ARCWHEAT, modified for CO_2 responses to illustrate potential production of winter wheat based on weather data from Rothamsted, England. The straight lines correspond to total dry matter production and grain yield predicted at 340 μ mol/mol CO_2 and observed weather. The curves are predicted production and yield at 680 μ mol/mol CO_2 and increases of daily mean temperature up to 4°C*

CO_2 and Water Availability

The influence of CO_2 on the water-use of crops may well prove to be the most important benefit of increased CO_2 concentrations for agriculture. Morison reviewed the recent literature and discussed the effect at a range of scales, pointing out that there are now many studies, including annual cereal crops, legumes and woody perennials, which show that, contrary to earlier assumptions, although the absolute amount of dry matter produced decreases at all CO_2 concentrations as water availability is reduced, the *relative* effect of increasing CO_2 on plant growth increases with decreasing water supply.

There are at least two mechanisms responsible for this relative enhancement: (1) the reduction in stomatal conductance, with a consequent beneficial effect on leaf water potential and hence on leaf expansion; (2) greater allocation of carbon to roots in elevated CO_2, and hence improved potential for exploring the soil volume for water and nutrients. In contrast, the increased production of leaf area in elevated CO_2 acts to increase the water-use per unit ground area of crops.

It is less clear whether increasing CO_2 will decrease water-use of crop canopies in the field over periods of days or months. This is not only because of the various mechanisms described above, but also because soil evaporation must be accounted for, and feedbacks between leaf temperature, atmospheric humidity and canopy evaporation apply. Consequently, the very rough estimate that a 33% increase in dry matter production combined with a 33% decrease in transpiration in response to doubling CO_2 concentration would increase water-use efficiency by a factor of 2 is unlikely to apply.

Goudriaan and Unsworth used a crop model with feedbacks between vegetation and the atmosphere to estimate that daily plant canopy water-use efficiency would increase by about 35% when carbon dioxide concentration doubled.

In a simpler model, but including soil evaporation, Morison calculated that seasonal water-use by crops with leaf area indices between 2 and 6 would decrease by only 10 to 15% in doubled CO_2 concentration. Nijs *et al.* (1989) grew ryegrass in 350 μ mol/mol CO_2 and 600 μ mol/mol CO_2. While the total water-use of the canopy was unaffected by growth in elevated CO_2, WUE was increased by 25% on a canopy basis or 87% on a leaf area basis. Results from this experiment indicated that, because the influence of elevated CO_2 on P_N and g_s was greatest early in the season, before the canopy closed, WUE probably varied seasonally, and this may be a general feature for annual crops.

Azam Ali (pers. comm.) used two methods to assess transpiration of crops of sorghum and groundnut growing at 350 and 700 μ mol/mol CO_2. Seasonal monitoring of soil water content with a neutron probe, combined with metred irrigation and subsidiary measurements of soil evaporation, enabled cumulative crop transpiration to be calculated. Porometry on individual leaves, and measurements of leaf area distribution, enabled transpiration per unit leaf area to be calculated. In groundnut but not in sorghum, elevated CO_2 reduced transpiration per unit leaf area throughout the season, but more so in the earlier stages.

On an even larger scale, de Bruin and Jacob estimated the influence of doubling CO_2 concentration on regional-scale transpiration, allowing for interactions between transpiration, leaf temperature and the planetary boundary layer. They concluded that, for vegetation where stomatal conductance decreased by 34% in response to elevated CO_2, but leaf area was unchanged, regional transpiration on a typical summer day would decrease by about 11 % for short crops such as cultivated grass and 17% for tall crops such as plantation forests. If negative feedback from the planetary boundary layer had been ignored, these values would have been overestimated (15 and 24%, respectively).

CO_2 and Ozone

When Krupa and Kickert attempted to review the interactive responses of crops to CO_2 and O_3 they found no publications describing experimental work. We summarize here the few papers that have been published since then, and comment on other work known to be in progress.

Allen made an estimate of the possible scale by which CO_2 might influence responses of soybeans to O_3. He used published results relating O_3 exposure to yield and a model of the sensitivity of stomatal conductance for soybean to CO_2. According to these models, doubling CO_2 concentration would reduce g_s by about 30% and thus decrease the stomatal uptake of O_3 by the crop. Allen estimated that this would result in a yield decrease of about 15% compared with yield in ambient air. This estimate takes no account of the large direct effects of CO_2

on soybean growth and assumes no interactions between the two gases. Scientists at the USDA Beltsville Research Centre have studied soybeans growing in open-top chambers in the field and exposed to CO_2 and O_3. Kramer *et al.* (1991), in a preliminary study, showed that the yield loss attributed to ozone at the site was about 12%, but when CO_2 concentrations were increased to 500 μ mol/mol the yield loss was only 6.7%. Mulchi *et al.* (1992) reported a more complex design with three O_3 treatments (charcoal-filtered, CF, non-filtered, NF, NF + 40 μ mol/mol O_3) and three CO_2 treatments (ambient, + 50, + 150 μ mol/mol). Increasing CO_2 concentration increased grain yield and grain oil content, and decreased protein content.

Increasing O_3 reduced grain yield but did not significantly alter oil or protein content. Table summarizes the interactive effects of CO_2 and O_3 on growth, yield and quality. Ozone caused a significant decrease in yield compared to the charcoal-filtered controls. When CO_2 was added to the ozone treatments, it partially counteracted the negative effect of O_3 so that yields in chambers with plus 150 μ mol/mol CO_2 and NF air were similar to those in ambient CO_2 and CF air. The results suggested that most effects of CO_2 and O_3 were additive.

Barnes and Pfirrman (1992) grew radishes in a phytotron in Munich, Germany, at two CO_2 concentrations (385 and 765 μ mol/mol) and two O_3 concentrations (20 nmol/mol and 73 nmol/mol), and studied gas exchange, growth and productivity, and mineral composition of the crop. In clean air the typical effects of elevated CO_2 on photosynthesis (increased) and stomatal conductance (decreased) were observed; the extra carbon assimilation stimulated root growth rate by 43% but there was no significant effect on shoot growth or leaf area. At ambient CO_2, the high O_3 treatment depressed photosynthesis by 26% (compared to the low treatment) and induced a slight reduction in g_s, with the net result that water-use efficiency declined. The reduction in carbon uptake was reflected in reduced growth, with roots more affected than shoots. Interactive effects of ozone and CO_2 were generally consistent with a reduction in ozone-induced responses in elevated CO_2. Early in the experiment, elevated CO_2 reduced the effect of O_3 on photosynthesis, but later this benefit disappeared, perhaps because long-term exposure to O_3 has an overall limiting effect on plant response to CO_2. The effects of CO_2 and O_3 on g_s appeared to be additive. Work is in progress in Munich on responses of grass swards to O_3, CO_2 and water availability. Preliminary results indicate that the growth depression which is observed as water is withheld in ambient CO_2 is alleviated at higher CO_2 levels. Responses of yield and water-use efficiency to the mixtures of CO_2 and ozone seem to be additive rather than synergistic.

In Newcastle, England, work is in progress exposing winter and spring cultivars of wheat to two CO_2 concentrations and two O_3 concentrations. Early results indicate that after 50 days' exposure there was evidence of downregulation of photosynthesis in response to prolonged exposure to elevated CO_2; this was most pronounced in the winter cultivars. Long-term exposure to O_3 resulted in a decrease in the light-saturated rate of CO_2 assimilation, partial stomatal closure, and accumulation of water soluble carbohydrate and starch in leaves. These resulted in decreased growth, with root growth more severely affected than shoot growth. In plants exposed to elevated CO_2 and O_3, effects on stomatal conductance were less than additive, and CO_2 enhancement effects on photosynthesis and growth were reduced by O_3.

In summary, from the very limited experimental results available at the time of writing, interactions between exposures to elevated O_3 and CO_2 seem to result in approximately additive effects on growth and yield. Both gases decrease stomatal conductance, and this results in less uptake of O_3 in elevated CO_2 treatments than in ambient air. Several authors have noted that this reduces (or at least delays) the damaging effects of O_3 In principle, the reverse is also true, i.e., increased O_3, by reducing stomatal conductance, negating some of the benefits to growth that elevated CO_2 confers. However, the O_3 exposure necessary to induce stomatal closure is generally sufficient to cause visible leaf injury, and this disbenefit is likely to be far more damaging than reduced CO_2 uptake through stomata. More field-based studies are necessary to assess fully the effects of the two gases in combination.

CO_2 and UV-B Radiation

Many studies of effects of UV-B radiation on crops are seriously flawed because UV exposures were either inadequately specified and/or were unrealistically large. Tevini (1993), in a thorough review of effects of UV-B radiation on plants, pointed out that, since the spectrum of UV from artificial light sources differs from that in the solar spectrum, and since many photobiological processes are strongly wavelength dependent, it is important to use weighting functions based on the action spectrum for specific responses before experiments in different exposure systems can be compared. Unfortunately, action spectra for many responses remain unknown.

In addition, many plant responses to UV-B are larger when plants are growing at low light levels typical of growth chambers and some glasshouses than they are in full sunshine, probably because natural protective pigments are inadequately synthesized and repair processes are restricted in low light. For this reason, field-based studies of crops are most relevant for estimating responses of productivity and yield to UV-B.

Even so, most field studies have used continuous UV-B exposures corresponding to 15 to 25% reductions in stratospheric O_3 i.e. roughly 30 to 50% increases in UV-B above present values. Such increases are much larger than are likely to occur as long-term means in agricultural regions under any likely depletion of stratospheric ozone in the next 30 years. Effects of UV-B on yield of crops growing in the field have produced very variable results. Work in the late 1970s and early 1980s by Biggs and colleagues in Florida, using very high UV exposures, found yield reductions in only about half the crops studied.

Work in Germany, also at high UV exposures, found no UV-B effects on cabbage, lettuce and rape. More recently, a large field-based study in Beltsville, Maryland, by Teramura, has revealed a high degree of intra-specific variability among soybean cultivars. The two most sensitive cultivars were grown for five seasons in the field in two UV treatments, corresponding to 16 and 25% ozone depletion simulations.

In one cultivar (Essex) yield was reduced by 20 to 25% at the high UV treatment; in the other cultivar, yield generally increased by 10 to 22% in this treatment. Neither cultivar showed a consistent significant yield change at the lower UV exposure. The conclusion from these studies must be that, under any realistic scenario of increasing UV-B radiation, crop yields are unlikely to be altered to a significant extent. We therefore turn to the question of whether UV-B exposure alters the sensitivity of crops to other atmospheric changes.

Effects of the combination of elevated CO_2 and UV-B radiation on crop growth and yield have been addressed in very few studies. Teramura and his colleagues grew wheat, rice and soybean in a glasshouse experiment. Treatments in the factorial design were:

ambient CO_2 (350 (m mol/mol), ambient UV, elevated CO_2 (650 μ mol/mol) and elevated UV-B (corresponding to a 10% decrease in stratospheric O_3 at the equator). Compared to the control, seed yield and total biomass increased significantly in elevated CO_2 for all three species. However, when UV-B and CO_2 were increased simultaneously, no increase in either seed yield (wheat and rice) or total biomass (rice) was observed compared to the control. In contrast, in soybean, the increases in seed yield and biomass induced by CO_2 alone were maintained in the high CO_2/high UV-B environment. Studies of leaf gas exchange indicated that UV-B reduced the apparent carboxylation efficiency in wheat and rice, but not in soybean.

Results of a CO_2/UV study on seedlings of sunflower and maize reported by Tevini (1993) are confounded by temperature changes in some treatments. A 25% increase in UV-B at 340 μ mol/mol CO_2 changed sunflower and maize dry weights (W) by -14 and -24%, respectively (compared to ambient controls). The same UV-B increment combined with a +2°C temperature change altered W by +5 % in sunflower and +31 % in maize. Adding a doubled CO_2 concentration to the elevated UV and temperature regime altered W by +19 and +32% for sunflower and maize, respectively. The incomplete experimental design limits interpretation of this study, but the results suggest that temperature and CO_2 are much more important influences on growth than any likely realistic increases in UV-B.

Ozone

There are a number of recent reviews of the mechanism of action of O_3 on plants, and of crop responses to O_3 (e.g., Tingey and Andersen, 1991; Heck *et al.*, 1988); consequently, we will only briefly summarize the main features. The response of plants to O_3 may be viewed as the culmination of a sequence of physical, biochemical and physiological events. The O_3 diffuses from the air into the leaf through stomata, and these exert control on O_3 uptake.

Plants are able to detoxify O_3 or its metabolites and can repair or compensate for O_3 impacts, so visible signs of ozone injury may not occur if the rate of O_3 uptake is sufficiently small. When resources have to be devoted to repair or compensation mechanisms, or when O_3 uptake is too large for full repair to be achieved, growth and yield may be reduced in the absence of visible injury by O_3. The principal modes of action of O_3 on plants are through injury to proteins and membranes, reduction in photosynthesis (i.e., carbon gained), changes in allocation of carbohydrate, and acceleration of senescence. The main result of O_3 exposure is therefore a reduction in the capacity of the plant to accumulate photosynthate through loss in photosynthetic capacity and increased foliage senescence.

Methods for exposing crops to ozone in experiments designed to develop yield/exposure responses are well established. The most commonly used technique is the open-top field chamber, which enables crop stands to be grown in either charcoal-filtered air or in air to which known amounts of O_3 have been added.

Coordinated programmes in the United States and Europe have investigated crop responses to O_3 in open-top chambers, and relationships between yield and seasonal mean O_3 concentration are available for many of the major world crops. The seasonal mean index, however, has been shown to be inadequate in relating ozone exposure to effects, primarily because of its inability to consider duration of exposure and its consideration of all concentrations as equal in their effect. Both duration and concentration are important in the effect of ozone (Hogsett *et al.,* 1988).

In particular, studies have demonstrated that ozone's effect is cumulative and that higher concentrations are more important than lower concentrations in causing an effect. Consequently, the cumulative, peak-weighted indices are more appropriate because of their biological relevancy. Examples of such indices include SUM06 (in units of nmol/mol-h) which cumulates all concentrations equal to or greater than 0.06 nmol/mol, or SIGMOID which cumulates all concentrations during the season and weights all concentrations according to a sigmoid function. The US Environmental Protection Agency (EPA) Draft Ozone Criteria Document (1995) reviewed and tabulated the latest information concerning effects of ozone on crop yields using both the seasonal mean, and two of the cumulative, peak-weighted indices.

The data for crop yield response are primarily from the National Crop Loss Assessment Network (NCLAN). Yield reductions in the NCLAN studies are calculated relative to a typical background O_3 concentration of SUM06 of O_3 or a 7-h seasonal mean of 25 nmol/mol. Yield losses of 10% or less in 50% of the crops (12 crop species, 54 studies including well-watered and drought studies) would occur with SUM06 exposure concentrations of 26.4 nmol/mol-h in three months, or 49 nmol/mol 7-h seasonal mean. Some species/cultivars were particularly sensitive; 11 % would be expected to have a yield reduction of 10% at a 7-hr seasonal mean of less than 35 nmol/mol. Similarly, 18% of the crops/cultivars are expected to have yield reductions of 10% or less at 3-month SUM06 concentrations of less than 10 nmol/mol-h. Both of these exposure values are quite low compared to the decade-average (1982-1991) exposure (3 months) from all monitoring sites across the United States (SUM06 = 29.5 nmol/mol-h; 7-h mean = 54 nmol/mol). In general, grain crops are less sensitive than others, but within-species variability and sensitivity may be greater than between species. Krupa and Kickert (1989) reviewed various published lists of the sensitivity of crop yields to O_3 onion, spinach, potato, alfalfa and cotton appear to be particularly sensitive crops, although it is possible to find resistant cultivars of all these species. Legumes range between sensitive and intermediate, and cereal and grass crops tend to be relatively resistant.

There is also a very large literature describing biotic and abiotic factors that modify plant response to 0. We will concentrate in this chapter on only a few interacting factors that are particularly relevant in terms of future changes in the atmospheric environment.

Ozone and Temperature

When interactions between ozone exposure and temperature were reviewed by the US EPA (1986) it was concluded that there were variable and conflicting results. More recent analysis (US EPA, 1995) suggests that many of these studies in controlled environments confounded effects of temperature with changes in vapour pressure deficit (VPD) because relative humidity was kept constant while temperature was increased. Sinco VPD can have profound effects on stomatal responses, evaporation rates and leaf expansion, it is not surprising that these early results have led to confusion. Recently, Todd *et al.* (1991) designed an experiment with tomato seedlings in which differences in VPD at different temperatures were minimized during O_3 exposure.

This study showed that, out of 11 growth variables measured, the only significant modifications by temperature of the effects of O_3 were on stem fresh weight and specific leaf area (leaf area/leaf dry weight). The authors suggested that VPD probably plays a more important role in determining sensitivity to O_3 than temperature. It is important to have more studies to clarify this issue, as many field-based studies of ozone responses use open-top chambers in which temperatures are increased by a few degrees above ambient. If either the changed temperature or the changed VPD in such chambers alters the sensitivity of the crops to O_3 there would be serious consequences for the application of much of our knowledge of the O_3 sensitivity of crops that has been gained from field-chamber experiments.

A second important O_3/temperature interaction, particularly affecting perennial species, including trees, is winter hardiness. Several studies have shown that exposures to O_3 reduce the frost-hardiness; Davison *et al.* (1988) reviewed the subject. Whilst most of this research has been with woody species, Barnes *et al.* (1988) showed that daily exposures to about 80 nmol/mol O_3 for 7 days significantly reduced the survival of pea seedlings after exposure to nighttime temperatures of about -4°C.

It seems likely that, as with sulphur dioxide, exposure to ozone could alter the frost hardiness of a number of agricultural crops. This response would be most likely to be significant in the event of an early autumn frost on early-sown crops such as winter cereals. There are suggestions that exposure of trees in summer to ozone alters the hardiness the following winter. Eamus and Murray (1991) pointed out that, even in severe winters, there are brief periods of mild temperatures that induce partial dehardening of woody species. They speculated that O_3 increases tree predisposition to dehardening in such conditions and consequently puts trees at greater risk from low temperatures. There are likely to be significant differences between species in this phenomenon. For example, in Florida, Eissenstat *et al.* (1991) found that, although O_3 reduced frost hardiness of citrus and avocado, the effects were small and the likelihood of significantly changing frost resistance is slight.

Ozone and Water Availability

The availability of soil water is one of the strongest influences on crop growth and productivity. When available, irrigation is commonly used to alleviate water stress, but in conditions of large evaporative demand, even irrigated crops may experience water stress. The most immediate effects of water stress are on plant-water potential, and may lead to reductions in leaf expansion rates and altered partitioning of dry matter, with increased assimilate partitioning to roots. When stomatal closure occurs, there is a resulting reduction in transpiration and CO_2 uptake and an increase in leaf temperature.

Water Stress and O_3 Responses

There has long been a general belief that water stress reduces the magnitude of any adverse effects of O_3 i.e., leaf injury, growth and yield reductions. This belief was based on observations that stomatal closure, induced by drought, reduced the rate and quantity of O_3 absorbed by leaves (US EPA, 1986).

However, more recent results show that the O_3/water availability interaction is complex, and probably depends on the magnitude and timing of the water stress (which has seldom been adequately defined in air pollution experiments). One difficulty in quantifying the interaction is the very large effect that water stress alone may have on yield, so that any influence of O_3 is sometimes masked.

Temple *et al.* (1985) studied the effects of O_3 and water stress on cotton growing in California. In this typically hot, dry season, the water-stressed plants wilted frequently, and yielded much less than the well-watered controls; nevertheless, when compared to yields of plants in a background O_3 concentration of 25 nmol/mol, and expressed as a percentage, the estimated yield loss at a seasonal mean 0, concentration of 50 nmol/mol was 7% in the well-watered plants and 2% in the water-stressed crop. Clearly the absolute yield loss from drought was much larger than the yield loss attributed to O_3 in either treatment.

Heagle *et al.* (1983) reviewed six studies of soybean responses to O_3 and water stress, and concluded that in only three were there significant interactions, i.e., the clear negative relationships between yield and O_3 exposure observed with well-watered plants were much reduced with water stressed plants. More recently, Heggestad and Lesser (1990) analysed three years of data for four soybean cultivars. They concluded that, in most cases, the relationships between yield and O_3 concentration had similar slopes for the water-stressed and well-watered treatments.

If this conclusion applies more generally, the extra yield losses which would result from O_3 exposure could be estimated if a yield response to drought alone is known. Clearly, there is still uncertainty over the influence of water stress on the form of the yield response to O_3. This uncertainty is unlikely to be resolved until better experimental designs with more precise specifications of the degree of water stress are developed.

O_3 and Water-use Efficiency

For irrigated crops where water supplies may be limited or expensive, it is important to know whether O_3 would influence the water-use efficiency (WUE) and consequently the amount of water required in a season. Reich *et al.* (1985) exposed well-watered soybeans to 130 nmol/mol O_3 for 7 hours daily, and found a 25% decrease in WUE compared with controls in 10 nmol/mol O_3.

Similar results were found for alfalfa (Temple *et al.*, 1988), probably indicating that carbon dioxide uptake was reduced more by O_3 than water loss. Tingey *et al.* (1994) also reported a significant decrease in WUE in soybeans exposed to episodic exposure regimes of 20-30 nmol/mol-h (3-month SUM06).

The study found the O_3 did not close stomata, but rather affected CO_2 assimilation. A significant increase in leaf construction cost was reported, indicating that both carbon and water are used inefficiently by plants exposed to ozone. In contrast, Greitner and Winner (1988) found that O_3 exposure *increased* the WUE of radish and soybeans. This disagreement probably reflects variation with exposure dynamics, stage of development, and genetic variation between and within species.

Recently Barnes and Pfirrman (1992) investigated effects of O_3 and elevated CO_2 singly and in combination on the WUE of radish. Both gases reduced stomatal conductance, and the combination of gases reduced conductance still further. O_3 reduced the instantaneous WUE, whereas CO_2 increased it (as discussed earlier). In the combination of O_3 and CO_2, there was initially no significant effect of O_3 (i.e., the response of WUE was similar to that observed in elevated CO_2), but as growth progressed, the WUE was significantly reduced in the combination treatment compared to in elevated CO_2 alone. This may be a response to accelerated senescence induced by O_3 exposure.

In summary, relatively severe water stress may reduce the yield losses attributable to O_3 in some crop species but, in general, the yield loss resulting from the water stress outweighs the benefits of the O_3 protection. For well-watered crops, O_3 exposure may change the water-use efficiency compared with crops in clean air, but the direction and magnitude of change probably depend on exposure and genetic factors. The influence of O_3 and CO_2 in combination on WUE is potentially important and merits further study.

Ozone and UV-B Radiation

Krupa and Kickert (1989) found no reports of O_3/UV-B interactions in their review. They assessed potential risks on a geographical basis, using distributions of major crops, tropospheric ozone concentrations and UV-B irradiance. They suggested that interactions might involve episodic exposure to the two stresses, so that peaks of O_3 would coincide with lower UV-B irradiance and vice versa. Runeckles and Krupa (1994) have developed this concept further, arguing that, as tropospheric O_3 increases, UV-B irradiance at the surface is reduced, because the O_3 absorbs some UV.

In theory, this effect could be important, in spite of the relatively small contribution of tropospheric O_3 to the total atmospheric O_3 column, because scattering by aerosols and molecules in the troposphere increases the radiation path length (Bruhl and Crutzen, 1989). In an attempt to quantify this effect, Albar (1992) compared the measured solar spectral irradiance I (1) at the ground near Nottingham, England, on two days when the tropospheric O_3 concentration was 51 and 84 nmol/mol, respectively, and stratospheric O_3 was constant. Figure shows that I (1) was about 20 to 40% greater in the UV-B waveband (280-320 nm) on the low O_3 day than on the high O_3 day, apparently supporting the hypothesis of Runeckles and Krupa (1994). However I (1) at longer wavelengths was also greater (by about 20 to 25%), and this increase is most probably a consequence of less aerosol (dust) being present on the low O_3 day. Since this aerosol effect would also apply in the UV-B waveband, it seems likely that the UV-B increase *directly* attributable to reduced tropospheric O_3 was no more than 10% to 15%. Variations in aerosol from day to day seem likely to be more important than variations in tropospheric O_3 in modulating the intensity of UV-B at the ground, but further observations would be valuable.

As mentioned earlier, many investigations of plant responses to UV-B are seriously flawed by inadequate specification and/or excessive UV exposure. An improved field-based system at Raleigh, North Carolina, using open-top chambers for studying crop responses to UV and O_3 has been used for three seasons of research on soybeans (Miller *et al.,* 1994). The system exposed soybean crops from emergence to maturity to UV treatments (ranging from ambient to about twice ambient biologically effective UV-B) and to O_3 treatments giving seasonal mean 12 h/d O_3 concentrations from 14 to 83 nmol/mol.

The ozone treatments resulted in reductions of photosynthesis, accelerated senescence, and reduced yield, in agreement with many other published studies. In contrast, the UV treatments, even at this relatively large UV irradiance, did not induce any significant changes in photosynthesis or yield, and there were no UV/O_3 interactions (Booker *et al.*, 1992b).

Fiscus *et al.* (1994) attempted to reconcile the lack of response to UV reported in the Raleigh experiments with other reports which indicate that increased UV-B causes physiological dysfunction and reductions in crop yields. They concluded that three factors make the conclusions of other field and glasshouse studies, at best, hard to interpret and possibly misleading: a frequent failure to monitor UV-B adequately; a tendency to underestimate UV exposures when relying on model calculations which do not allow sufficiently for effects of dust in the atmosphere; and no adjustment of UV exposure for seasonal and daily weather changes (and hence a tendency towards unrealistically large exposures).

Although it would be useful to see further well-designed field studies of UV/O_3 interactions using a wide range of crop and natural species, on the evidence of the careful work of Booker, Miller, Fiscus and their colleagues, it seems unlikely that UV/O_3 interactions are of any importance for crop productivity, and it seems clear that O_3 poses a much greater threat to yields of many crops than any likely increases in UV-B radiation.

3

Famine and its Effects on Agriculture

A famine is a widespread scarcity of food, caused by several factors including crop failure, population unbalance, or government policies. This phenomenon is usually accompanied or followed by regional malnutrition, starvation, epidemic, and increased mortality. Nearly every continent in the world has experienced a period of famine throughout history. Some countries, particularly in sub-Sahara Africa, continue to have extreme cases of famine.

The famine relief model increasingly used by aid groups calls for giving cash or cash vouchers to the hungry to pay local farmers instead of buying food from donor countries, as is often required by law (for example U.S. law requires that food aid money be spent on food grown in the U.S.), as it wastes money on transport costs, but more importantly, it perpetuates the cycle of dependency on foreign imports rather than helping to create real local stability through agricultural abundance. Emergency measures in relieving famine include providing high calorie ready-to-use therapeutic food (RUTF), through fortified sachets of peanut-based paste such as Plumpy'nut that are given primarily to children.

Long-term measures include investment in modern agriculture techniques, such as fertilizers and irrigation, which largely eradicated hunger in the developed world. World Bank strictures restrict government subsidies for farmers, and increasing use of fertilizers is opposed by some environmental groups because of its unintended consequences: adverse effects on water supplies and habitat.

Characteristics

Famine strikes Sub-Saharan African countries the hardest, but with exhaustion of food resources, overdrafting of groundwater, wars, internal struggles, and economic failure, famine continues to be a worldwide problem with hundreds of millions of people suffering. These famines cause widespread malnutrition and impoverishment; The famine in Ethiopia in the 1980s had an immense death toll, although Asian famines of the 20th century have also produced extensive death tolls. Modern African famines are characterized by widespread destitution and malnutrition, with heightened mortality confined to young children.

Relief technologies including immunization, improved public health infrastructure, general food rations and supplementary feeding for vulnerable children, has provided temporary mitigation to the mortality impacts of famines, while leaving their economic consequences unchanged, and not solving the underlying issue of too large a regional population relative to food production capability. Humanitarian crises may also arise from genocide campaigns, civil wars, refugee flows and episodes of extreme violence and state collapse, creating famine conditions among the affected populations.

Despite repeated stated intentions by the world's leaders to end hunger and famine, famine remains a chronic threat in much of Africa and Asia. In July 2005, the Famine Early Warning Systems Network labelled Niger with emergency status, as well as Chad, Ethiopia, South Sudan, Somalia and Zimbabwe. In January 2006, the United Nations Food and Agriculture Organization warned that 11 million people in Somalia, Kenya, Djibouti and Ethiopia were in danger of starvation due to the combination of severe drought and military conflicts. In 2006, the most serious humanitarian crisis in Africa was in Sudan's region Darfur.

Some believed that the Green Revolution was an answer to famine in the 1970s and 1980s. The Green Revolution began in the 20th century with hybrid strains of high-yielding crops. Between 1950 and 1984, as the Green Revolution transformed agriculture around the globe, world grain production increased by 250%. Some criticize the process, stating that these new high-yielding crops require more chemical fertilizers and pesticides, which can harm the environment. However, it was an option for developing nations suffering from famine. These high-yielding crops make it technically possible to feed more people. However, there are indications that regional food production has peaked in many world sectors, due to certain strategies associated with intensive agriculture such as groundwater overdrafting and overuse of pesticides and other agricultural chemicals.

Frances Moore Lappé, later co-founder of the Institute for Food and Development Policy (Food First) argued in *Diet for a Small Planet* (1971) that vegetarian diets can provide food for larger populations, with the same resources, compared to omnivorous diets.

Noting that modern famines are sometimes aggravated by misguided economic policies, political design to impoverish or marginalize certain populations, or acts of war, political economists have investigated the political conditions under which famine is prevented. Economist Amartya Sen states that the liberal institutions that exist in India, including competitive elections and a free press, have played a major role in preventing famine in that country since independence. Alex de Waal has developed this theory to focus on the "political contract" between rulers and people that ensures famine prevention, noting the rarity of such political contracts in Africa, and the danger that international relief agencies will undermine such contracts through removing the locus of accountability for famines from national governments.

Effects

The demographic impacts of famine are sharp. Mortality is concentrated among children and the elderly. A consistent demographic fact is that in all recorded famines, male mortality exceeds female, even in those populations (such as northern India and Pakistan) where there is a male longevity advantage during normal times. Reasons for this may include greater female resilience under the pressure of malnutrition, and possibly female's naturally higher percentage of body fat. Famine is also accompanied by lower fertility. Famines therefore leave the reproductive core of a population—adult women—lesser affected compared to other population categories, and post-famine periods are often characterized a "rebound" with increased births. Even though the theories of Thomas Malthus would predict that famines reduce the size of the population commensurate with available food resources, in fact even the most severe famines have rarely dented population growth for more than a few years. The mortality in China in 1958–61, Bengal in 1943, and Ethiopia in 1983–85 was all made up by a growing population over just a few years. Of greater long-term demographic impact is emigration: Ireland was chiefly depopulated after the 1840s famines by waves of emigration.

Levels of Food Insecurity

In modern times, local and political governments and non-governmental organizations that deliver famine relief have limited resources with which to address the multiple situations of food insecurity that are occurring simultaneously. Various methods of categorizing the gradations of food security have thus been used in order to most efficiently allocate food relief. One of the earliest were the Indian Famine Codes devised by the British in the 1880s. The Codes listed three stages of food insecurity: near-scarcity, scarcity and famine, and were highly influential in the creation of subsequent famine warning or measurement systems. The early warning system developed to monitor the region inhabited by the Turkana people in northern Kenya also has three levels, but links each stage to a pre-planned response to mitigate the crisis and prevent its deterioration

The experiences of famine relief organizations throughout the world over the 1980s and 1990s resulted in at least two major developments: the "livelihoods approach" and the increased use of nutrition indicators to determine the severity of a crisis. Individuals and groups in food stressful situations will attempt to cope by rationing consumption, finding alternative means to supplement income, etc. before taking desperate measures, such as selling off plots of agricultural land. When all means of self-support are exhausted, the affected population begins to migrate in search of food or fall victim to outright mass starvation. Famine may thus be viewed partially as a social phenomenon, involving markets, the price of food, and social support structures. A second lesson drawn was the increased use of rapid nutrition assessments, in particular of children, to give a quantitative measure of the famine's severity.

Since 2003, many of the most important organizations in famine relief, such as the World Food Programme and the U.S. Agency for International Development, have adopted a five-level scale measuring intensity and magnitude. The intensity scale uses both livelihoods' measures and measurements of mortality and child malnutrition to categorize a situation as food secure, food insecure, food crisis, famine, severe famine, and extreme famine. The number of deaths determines the magnitude designation, with under 1000 fatalities defining a "minor famine" and a "catastrophic famine" resulting in over 1,000,000 deaths.

Causes

Definitions of famines are based on three different categories – these include food supply-based, food consumption-based and mortality-based definitions. Some definitions of famines are:

- Blix – Widespread food shortage leading to significant rise in regional death rates.
- Brown and Eckholm – Sudden, sharp reduction in food supply resulting in widespread hunger.
- Scrimshaw – Sudden collapse in level of food consumption of large numbers of people.
- Ravallion – Unusually high mortality with unusually severe threat to food intake of some segments of a population.
- Cuny – A set of conditions that occurs when large numbers of people in a region cannot obtain sufficient food, resulting in widespread, acute malnutrition.

Food shortages in a population are caused either by a lack of food or by difficulties in food distribution; it may be worsened by natural climate fluctuations and by extreme political conditions related to oppressive government or warfare. One of the proportionally largest historical famines was the Bengal Famine of 1770 in the lower Gangetic plain of East India Company ruled North East India. It began in 1770 due to severe extractive practices of the East India Company. An estimated ten million people died in the famine, roughly one in three people in the affected area. The deaths were greatly exacerbated by the fact that the East India Company raised land taxes by 10% at the height of the famine, in April 1770.

The conventional explanation until 1981 for the cause of famines was the Food availability decline (FAD) hypothesis. The assumption was that the central cause of all famines was a decline in food availability. However, FAD could not explain why only a certain section of the population such as the agricultural laborer was affected by famines while others were insulated from famines. Based on the studies of some recent famines, the decisive role of FAD has been questioned and it has been suggested that the causal mechanism for precipitating starvation includes many variables other than just decline of food availability. According to this view, famines are a result of entitlements, the theory being proposed is called the "failure of exchange entitlements" or FEE. A person may own various commodities that can be exchanged in a market economy for the other commodities he or she needs. The exchange can happen via trading or production or through a combination of the two. These entitlements are called trade-based or production-based entitlements. Per this proposed view, famines are precipitated due to a breakdown in the ability of the person to exchange his entitlements. An example of famines due to FEE is the inability of an agricultural laborer to exchange his primary entitlement, i.e., labour for rice when his employment became erratic or was completely eliminated.

According to the Physicians for Social Responsibility (PSR), global climate change is additionally challenging the Earth's ability to produce food, potentially leading to famine.

Some elements make a particular region more vulnerable to famine. These include poverty, population growth, an inappropriate social infrastructure, a suppressive political regime, and a weak or under-prepared government.

State-sponsored Famine

In certain cases, such as the Great Leap Forward in China (which produced the largest famine in absolute numbers), North Korea in the mid-1990s, or Zimbabwe in the early-2000s, famine can occur because of government policy. Few historians have argued that the Great Irish Famine was caused by the shortage of food, given that Ireland was producing enough food to feed its eight million people, but by the British government's choice to leave open the ports, as they are normally closed during Irish crop blights. Records show that in past famines, ports were closed to keep Irish-grown food in Ireland to feed the Irish. However, this did not occur in the 1840s, and Ireland continued to export food during the Famine. Law professor Charles E. Rice of Notre Dame as well as International Law professor at the University of Illinois Francis A. Boyle have argued that the British government committed genocide by pursuing a policy of starvation in Ireland. This claim has been debated throughout history, but also shows the troubles between the two countries.

In 1932, under the USSR, Ukraine experienced one of their largest famine as a result of state sponsored famine. Termed Holodomor, meaning killing by hunger, between 2.4 and 7.5 millions peasants died from a planned repression to eliminate protesters of collectivization as well. Forced grain quotas imposed upon the rural peasants and brutal terror contributed to the widespread famine despite the large grain harvest and good weather. Soviets continued to deny the problem and did not provide for victims nor accept foreign aid.

In 1958 in China, Mao Zedong's Communist Government began the Great Leap Forward campaign, aimed at rapidly industrializing the country. The government forcibly took control of agriculture. Barely enough grain was left for the peasants, and starvation set in many rural areas. Exportation of grain continued despite the famine to conceal the problem. While the famine is attributed to unintended consequences, it is believed that the government refused to acknowledge the problem, thereby further contributing to the deaths. In many instances, peasants were persecuted. Between 20 to 45 million people perished in this famine, making it one of the most deadly famines to date.

Malawi ended its famine by subsidizing farmers against the strictures of the World Bank. During the 1973 Wollo Famine in Ethiopia, food was shipped out of Wollo to the capital city of Addis Ababa, where it could command higher prices. In the late-1970s and early-1980s, residents of the dictatorships of Ethiopia and Sudan suffered massive famines, but the democracy of Botswana avoided them, despite also having a severe drop in national food production. In Somalia, famine occurred because of a failed state.

Many famines are caused by imbalance of food production compared to the large populations of countries whose population exceeds the regional carrying capacity. Historically, famines have occurred from agricultural problems such as drought, crop failure, or pestilence. Changing weather patterns, the ineffectiveness of medieval governments in dealing with crises, wars, and epidemic diseases such as the Black Death helped to cause hundreds of famines in Europe during the Middle Ages, including 95 in Britain and 75 in France. In France, the Hundred Years' War, crop failures and epidemics reduced the population by two-thirds.

The failure of a harvest or change in conditions, such as drought, can create a situation whereby large numbers of people continue to live where the carrying capacity of the land has temporarily dropped radically. Famine is often associated with subsistence agriculture. The total absence of agriculture in an economically strong area does not cause famine; Arizona and other wealthy regions import the vast majority of their food, since such regions produce sufficient economic goods for trade.

Famines have also been caused by volcanism. The 1815 eruption of the Mount Tambora volcano in Indonesia caused crop failures and famines worldwide and caused the worst famine of the 19th century. The current consensus of the scientific community is that the aerosols and dust released into the upper atmosphere causes cooler temperatures by preventing the sun's energy from reaching the ground. The same mechanism is theorized to be caused by very large meteorite impacts to the extent of causing mass extinctions.

Risk of Future Famine

The Guardian reports that in 2007 approximately 40% of the world's agricultural land is seriously degraded. If current trends of soil degradation continue in Africa, the continent might be able to feed just 25% of its population by 2025, according to UNU's Ghana-based Institute for Natural Resources in Africa. As of late 2007, increased farming for use in biofuels, along with world oil prices at nearly $100 a barrel, has pushed up the price of grain used to feed poultry and dairy cows and other cattle, causing higher prices of wheat (up 58%), soybean (up 32%), and maize (up 11%) over the year. In 2007 Food riots have taken place in many countries across the world. An epidemic of stem rust, which is destructive to wheat and is caused by race Ug99, has in 2007 spread across Africa and into Asia.

Beginning in the 20th century, nitrogen fertilizers, new pesticides, desert farming, and other agricultural technologies began to be used to increase food production, in part to combat famine. Between 1950 and 1984, as the Green Revolution influenced agriculture, world grain production increased by 250%. Much of this gain is non-sustainable. Such agricultural technologies temporarily increased crop yields. However, as early as 1995, there were signs that they may be contributing to the decline of arable land (e.g. persistence of pesticides leading to soil contamination and decline of area available for farming). Developed nations have shared these technologies with developing nations with a famine problem, but there are ethical limits to pushing such technologies on lesser developed countries. This is often attributed to an association of inorganic fertilizers and pesticides with a lack of sustainability.

David Pimentel, professor of ecology and agriculture at Cornell University, and Mario Giampietro, senior researcher at the National Research Institute on Food and Nutrition (INRAN), place in their study *Food, Land, Population and the U.S. Economy* the maximum U.S. population for a sustainable economy at 200 million. To achieve a sustainable economy and avert disaster, the United States must reduce its population by at least one-third, and world population will have to be reduced by two-thirds, says study. The authors of this study believe that the mentioned agricultural crisis will only begin to impact us after 2020, and will not become critical until 2050. The oncoming peaking of global oil production (and subsequent decline of production), along with the peak of North American natural gas production will very likely precipitate this agricultural crisis much sooner than expected.

Geologist Dale Allen Pfeiffer claims that coming decades could see spiraling food prices without relief and massive starvation on a global level such as never experienced before. Water deficits, which are already spurring heavy grain imports in numerous smaller countries, may soon do the same in larger countries, such as China or India. The water tables are falling in scores of countries (including Northern China, the US, and India) due to widespread overpumping using powerful diesel and electric pumps. Other countries affected include Pakistan, Iran, and Mexico. This will eventually lead to water scarcity and cutbacks in grain harvest. Even with the overpumping of its aquifers, China has developed a grain deficit, contributing to the upward pressure on grain prices. Most of the three billion people projected to be added worldwide by mid-century will be born in countries already experiencing water shortages.

After China and India, there is a second tier of smaller countries with large water deficits — Algeria, Egypt, Iran, Mexico, and Pakistan. Four of these already import a large share of their grain. Only Pakistan remains marginally self-sufficient. But with a population expanding by 4 million a year, it will also soon turn to the world market for grain. According to a UN climate report, the Himalayan glaciers that are the principal dry-season water sources of Asia's biggest rivers - Ganges, Indus, Brahmaputra, Yangtze, Mekong, Salween and Yellow - could disappear by 2350 as temperatures rise and human demand rises. Approximately 2.4 billion people live in the drainage basin of the Himalayan rivers. India, China, Pakistan, Afghanistan, Bangladesh, Nepal and Myanmar could experience floods followed by severe droughts in coming decades. In India alone, the Ganges provides water for drinking and farming for more than 500 million people.

Famine Action

Prevention: The effort to bring modern agricultural techniques found in the Western world, such as nitrogen fertilizers and pesticides, to Asia, called the Green Revolution, resulted in decreases in malnutrition similar to those seen earlier in Western nations. This was possible because of existing infrastructure and institutions that are in short supply in Africa, such as a system of roads or public seed companies that made seeds available. Supporting farmers in areas of food insecurity, through such measures as free or subsidized fertilizers and seeds, increases food harvest and reduces food prices.

The World Bank and some rich nations press nations that depend on them for aid to cut back or eliminate subsidized agricultural inputs such as fertilizer, in the name of privatization even as the United States and Europe extensively subsidized their own farmers. Many, if not most, of the farmers are too poor to afford fertilizer at market prices. For example, in the case of Malawi, almost five million of its 13 million people used to need emergency food aid. However, after the government changed policy and subsidies for fertilizer and seed were introduced, farmers produced record-breaking corn harvests in 2006 and 2007 as production leaped to 3.4 million in 2007 from 1.2 million in 2005. This lowered food prices and increased wages for farm workers. Malawi became a major food exporter, selling more corn to the World Food Programme and the United Nations than any other country in Southern Africa. Proponents for helping the farmers includes the economist Jeffrey Sachs, who has championed the idea that wealthy countries should invest in fertilizer and seed for Africa's farmers.

Relief

Deficient micronutrients can be provided through fortifying foods. Fortifying foods such as peanut butter sachets and Spirulina have revolutionized emergency feeding in humanitarian emergencies because they can be eaten directly from the packet, do not require refrigeration or mixing with scarce clean water, can be stored for years and, vitally, can be absorbed by extremely ill children. The United Nations World Food Conference of 1974 declared Spirulina as 'the best food for the future' and its ready harvest every 24 hours make it a potent tool to eradicate malnutrition.

What is recommended by WHO and other sources for malnourished children or adults who also have diarrhea is drinking rehydration solution, continuing to eat, antibiotics, and zinc supplements. There is a special oral rehydration solution called ReSoMal which has less sodium and more potassium than standard solution. However, if the diarrhea is severe, the standard solution is preferable as the person needs the extra sodium. Obviously, this is a judgement call best made by a physician, and using either solution is better than doing nothing. Zinc supplements often can help reduce the duration and severity of diarrhea, and Vitamin A can also be helpful. The World Health Organization is quite emphatic on the importance of a person with diarrhea continuing to eat, with a 2005 publication for physicians stating: "Food should *never* be withheld and the child's usual foods should *not* be diluted. Breastfeeding should *always* be continued."

There is a growing realization among aid groups that giving cash or cash vouchers instead of food is a cheaper, faster, and more efficient way to deliver help to the hungry, particularly in areas where food is available but unaffordable. The United Nations's World Food Programme (WFP), the biggest non-governmental distributor of food, announced that it will begin distributing cash and vouchers instead of food in some areas, which Josette Sheeran, the WFP's executive director, described as a "revolution" in food aid. The aid agency Concern Worldwide is piloting a method through a mobile phone operator, Safaricom, which runs a money transfer programme that allows cash to be sent from one part of the country to another.

However, for people in a drought living a long way from and with limited access to markets, delivering food may be the most appropriate way to help. Fred Cuny stated that "the chances of saving lives at the outset of a relief operation are greatly reduced when food is imported. By the time it arrives in the country and gets to people, many will have died." US Law, which requires buying food at home rather than where the hungry live, is inefficient because approximately half of what is spent goes for transport. Fred Cuny further pointed out "studies of every recent famine have shown that food was available in-country — though not always in the immediate food deficit area" and "even though by local standards the prices are too high for the poor to purchase it, it would usually be cheaper for a donor to buy the hoarded food at the inflated price than to import it from abroad."

Ethiopia has been pioneering a programme that has now become part of the World Bank's prescribed recipe for coping with a food crisis and had been seen by aid organizations as a model of how to best help hungry nations. Through the country's main food assistance programme, the Productive Safety Net Programme, Ethiopia has been giving rural residents who are chronically short of food, a chance to work for food or cash. Foreign aid organizations like the World Food Programme were then able to buy food locally from surplus areas to distribute in areas with a shortage of food.

History

During the 20th century, an estimated 70 million people died from famines across the world, of whom an estimated 30 million died during the famine of 1958–61 in China. The other most notable famines of the century included the 1942–1945 disaster in Bengal, famines in China in 1928 and 1942, and a sequence of famines in the Soviet Union, including the Soviet famine of 1932-1933, Stalin's famine inflicted on USSR in 1932–33. A few of the great famines of the late 20th century were: the Biafran famine in the 1960s, the Khmer Rouge-caused famine in Cambodia in the 1970s, the Ethiopian famine of 1984–85 and the North Korean famine of the 1990s.

Africa

In the mid-22nd century BC, a sudden and short-lived climatic change that caused reduced rainfall resulted in several decades of drought in Upper Egypt. The resulting famine and civil strife is believed to have been a major cause of the collapse of the Old Kingdom. An account from the First Intermediate Period states, "All of Upper Egypt was dying of hunger and people were eating their children." In 1680s, famine extended across the entire Sahel, and in 1738 half the population of Timbuktu died of famine. In Egypt, between 1687 and 1731, there were six famines. The famine that afflicted Egypt in 1784 cost it roughly one-sixth of its population. At the end of the 18th century, and even more at the beginning of the nineteenth, the Maghreb suffered from the deadly combination of plague and famine. Tripoli and Tunis experienced famine in 1784 and 1785 respectively.

According to John Iliffe, "Portuguese records of Angola from the 16th century show that a great famine occurred on average every seventy years; accompanied by epidemic disease, it might kill one-third or one-half of the population, destroying the demographic growth of a generation and forcing colonists back into the river valleys."

With famine reigning over Africa, corruption in governments and foreign aid started to become more prevalent. National governments have control over each of these to varying degrees, although their implementation varies regionally. On the other hand, donors and corporations sometimes build infrastructures and give loans such as the world banks and NATO, but the continued protection and aid depends on the governments and agencies. Onc common theme in foreign aid and how it is disrupted is a correlation between geography and food security along with class, ethnicity, gender and religion. Foreign aid tends to find its way into hands that don't need it in corrupt governments, siphoning it away from the poor. Taking the aid away from the people that really need it slows the pace of development and progress ultimately undermines the welfare of citizens.

The first documentation of weather in West-Central Africa occurs around the mid-16th to 17th centuries in areas such as Luanda Kongo, however, not much data was recorded on the issues of weather and disease except for a few notable documents. The only records obtained are of violence between Portuguese and Africans during the battle of mbilwa in 1665. In these documents the Portuguese wrote of African raids on Portuguese merchants solely for food, giving clear signs of famine. Additionally, instances of cannibalism by the African Jaga were also more prevalent during this time frame, indicating an extreme deprivation of a primary food source.

Historians of African famine have documented repeated famines in Ethiopia. Possibly the worst episode occurred in 1888 and succeeding years, as the rinderpest epizootic, introduced into Eritrea by infected cattle, spread southwards reaching ultimately as far as South Africa. In Ethiopia it was estimated that as much as 90 percent of the national herd died, rendering rich farmers and herders destitute overnight. This coincided with drought associated with an el Nino oscillation, human epidemics of smallpox, and in several countries, intense war. The Ethiopian Great famine that afflicted Ethiopia from 1888 to 1892 cost it roughly one-third of its population. In Sudan the year 1888 is remembered as the worst famine in history, on account of these factors and also the exactions imposed by the Mahdist state.

Much of the famines occurring in the 19th and early 20th centuries were caused by wars involving colonization by the British, Spanish, Portuguese, and Belgian empires. Greed and the will to control the territory was what spurred most of these initial droughts into out of control famines. By withholding food supplies, the ensuing famine would suppress any rebellions or opposition to colonization. A prominent figure in the starvation of many Africans around the turn of the 20th century was Belgian King Leopold II who is credited with the forming of the Congo Free State.

In forming this state, Leopold used mass labour camps to finance his empire. These camps were occupied by enslaved Africans who were often starved to death. His empire contributed to the death of roughly 10 million Africans and has had long lasting effects on the current Democratic Republic of Congo in which famine and rape are still occurring. William Rubinstein wrote: "More basically, it appears almost certain that the population figures given by Hochschild are inaccurate. There is, of course, no way of ascertaining the population of the Congo before the twentieth century, and estimates like 20 million are purely guesses. Most of the interior of the Congo was literally unexplored if not inaccessible."

Colonial "pacification" efforts often caused severe famine, as for example with the repression of the Maji Maji revolt in Tanganyika in 1906. The introduction of cash crops such as cotton, and forcible measures to impel farmers to grow these crops, also impoverished the peasantry in many areas, such as northern Nigeria, contributing to greater vulnerability to famine when severe drought struck in 1913.

Another example of this pacification can be taken from the years revolving around 1896 in which the British empire destroyed large amounts of grain and food stocks, suppressing a riot being formed by the Ndeblee tribe.

Records compiled for the Himba recall two droughts from 1910-1917. They were recorded by the Himba through a method of oral tradition. From 1910-1911 the Himba described the drought as "drought of the omutati seed" also called *omangowi*, which means the fruit of an unidentified vine that people ate during the time period. From 1914-1916 droughts brought *katur' ombanda* or *kari' ombanda* which means "the time of eating clothing".

For the middle part of the 20th century, agriculturalists, economists and geographers did not consider Africa to be famine prone (they were much more concerned about Asia). There were notable counter-examples, such as the famine in Rwanda during World War II and the Malawi famine of 1949, but most famines were localized and brief food shortages. Although the drought was brief the main cause of death in Rwanda was due to Belgian prerogatives to acquisition grain from their colony (Rwanda). This and the drought caused 300,000 Rwandans to perish.

From 1967-1969 large scale famine occurs in Biafra and Nigeria due to the government blockading the Breakaway territory. It is estimated that 1.5 million people died of starvation due to this famine. Additionally, with the added effect of the drought and other corrupt governments withholding food during this time; 500,000 Africans perish in Central and West Africa.

The specter of famine recurred only in the early 1970s, when Ethiopia and the west African Sahel suffered drought and famine. The Ethiopian famine of that time was closely linked to the crisis of feudalism in that country, and in due course helped to bring about the downfall of the Emperor Haile Selassie. The Sahelian famine was associated with the slowly growing crisis of pastoralism in Africa, which has seen livestock herding decline as a viable way of life over the last two generations.

Since then, African famines become more frequent, more widespread and more severe until the start of the 21st century. Since then, more effective early warning and humanitarian response actions have reduced the number of deaths by famine markedly. That said, many African countries are not self-sufficient in food production, relying on income from cash crops to import food. Agriculture in Africa is susceptible to climatic fluctuations, especially droughts which can reduce the amount of food produced locally. Other agricultural problems include soil infertility, land degradation and erosion, swarms of desert locusts, which can destroy whole crops, and livestock diseases. The Sahara reportedly spreads at a rate of up to 30 miles a year. The most serious famines have been caused by a combination of drought, misguided economic policies, and conflict. The 1983–85 famine in Ethiopia, for example, was the outcome of all these three factors, made worse by the Communist government's censorship of the emerging crisis. In Sudan at the same date, drought and economic crisis combined with denials of any food shortage by the then-government of President Gaafar Nimeiry, to create a crisis that killed perhaps 250,000 people—and helped bring about a popular uprising that overthrew Nimeiry.

Numerous factors make the food security situation in Africa tenuous, including political instability, armed conflict and civil war, corruption and mismanagement in handling food supplies, and trade policies that harm African agriculture. An example of a famine created by human rights abuses is the 1998 Sudan famine. AIDS is also having long-term economic effects on agriculture by reducing the available workforce, and is creating new vulnerabilities to famine by overburdening poor households. On the other hand, in the modern history of Africa on quite a few occasions famines acted as a major source of acute political instability. In Africa, if current trends of population growth and soil degradation continue, the continent might be able to feed just 25% of its population by 2025, according to United Nations University (UNU)'s Ghana-based Institute for Natural Resources in Africa.

Recent examples include Sahel drought of the 1970s, Ethiopia in 1973 and mid-1980s, Sudan in the late-1970s and again in 1990 and 1998. The 1980 famine in Karamoja, Uganda was, in terms of mortality rates, one of the worst in history. 21% of the population died, including 60% of the infants.

In the 1980s there was large scale multilayer drought that occurred in the Sudan and Sahelian region of Africa. This was a large concern because even though the Sudanese Government believed there was a surplus of grain, there were local deficits that were occurring across the region.

In October 1984, television reports around the world carried footage of starving Ethiopians whose plight was centred around a feeding station near the town of Korem. BBC newsreader Michael Buerk gave moving commentary of the tragedy on 23 October 1984, which he described as a "biblical famine". This prompted the Band Aid single, which was organized by Bob Geldof and featured more than 20 pop stars. The Live Aid concerts in London and Philadelphia raised even more funds for the cause. An estimated 900,000 people died within one year as a result of the famine, but the tens of millions of pounds raised by Band Aid and Live Aid are widely believed to have saved the lives of Ethiopians who were in danger of dying.

A central reason as to why the famine (one of the largest seen in the country) is thought to have occurred is that Ethiopia (and the surrounding Horn) was still recovering from the droughts which occurred in the mid-late 1970s.

Compounding this problem was the intermittent fighting erupting in the region due to a civil war occurring in Ethiopia. Further issues were created by the government's lack of organization in providing relief; many even hoarded supplies as a way to control the population. Ultimately, over 1 million Ethiopians died and over 22 million people suffered due to the prolonged drought, which lasted roughly 2 years.

Subsequently in 1992 Somalia becomes a war zone with no effective government, police, or basic services. This is mainly due to the collapse of the dictatorship led by [Barre] and the split of power between warlords. Additionally a massive drought occurs at the same instance causing over 300,000 Somalians to perish.

Cases Since 2000

The 2005–06 Niger food crisis was a severe but localized food security crisis in the regions of northern Maradi, Tahoua, Tillabéri, and Zinder of Niger.

It was caused by an early end to the 2004 rains, desert locust damage to some pasture lands, high food prices, and chronic poverty. In the affected area, 2.4 million of 3.6 million people are considered highly vulnerable to food insecurity An international assessment stated that, of these, over 800,000 face extreme food insecurity and another 800,000 in moderately insecure food situations are in need of aid.

2010 Sahel Famine

The 2010 Sahel famine hit millions in Niger and across West Africa face food shortages after erratic rains hit farming in countries in the Sahel region south of the Sahara desert, the European Commission's aid group said Thursday.

The erratic rains in the 2009/2010 agricultural season have resulted in an enormous deficit in food production in these countries," he said of nations such as Niger, Chad, northern Burkina Faso and northern Nigeria. He said strong leadership would be required from the United Nations and the rest of the international community to mobilise aid. "If we work fast enough, early enough, it won't be a famine. If we don't there is a strong risk."

2011 East Africa Drought

In mid-2011, two consecutive missed rainy seasons precipitated the worst drought in East Africa in 60 years. The United Nations officially declared a famine in several regions in July of the year, a situation reportedly exacerbated by a temporary ban on relief supplies imposed by Islamist militants. An estimated 50,000 to 150,000 people are reported to have died during the period, though these figures and the extent of the crisis are disputed. In February 2012, the UN announced that the food crisis was over due to a scaling up of relief efforts and a bumper harvest. Aid agencies subsequently shifted their emphasis to recovery efforts, including digging irrigation canals and distributing plant seeds.

2012 Sahel Drought

July 6 saw the Methodist Relief and Development Fund (MRDF) aid experts say that more than 1,500,000 Nigerians were at risk of famine due to a month long heat wave that was hovering over Niger, Mali, Mauritania and Morocco. A fund of about £20,000 was distributed to the crisis-hit countries of Niger, Mali, Burkina Faso and Mauritania.

Initiatives to Increase Food Security

Against a backdrop of conventional interventions through the state or markets, alternative initiatives have been pioneered to address the problem of food security. An example is the "Community Area-Based Development Approach" to agricultural development ("CABDA"), an NGO programme with the objective of providing an alternative approach to increasing food security in Africa. CABDA proceeds through specific areas of intervention such as the introduction of drought-resistant crops and new methods of food production such as agro-forestry. Piloted in Ethiopia in the 1990s it has spread to Malawi, Uganda, Eritrea and Kenya. In an analysis of the programme by the Overseas Development Institute, CABDA's focus on individual and community capacity-building is highlighted. This enables farmers to influence and drive their own development through community-run institutions, bringing food security to their household and region.

Asia

Cambodia: In 1975, the Khmer Rouge entered the capital of Phnom Penh and took control of Cambodia. With the application of the fundamental ideals of communism, the new government under Pol

Pot drove all urban residents into the countryside to work on communal farm and civil work projects. Without external assistance, with 75% of the necessary draft animals dead from the previous four years of war, agricultural guidelines written by idealists, and work overseen by zealous cadre, the country soon sunk into the depths of famine. No international relief would come until the Vietnamese army invaded in 1979 and liberated the country. While Pol Pot was in power, between one and three million people died out of a total population of eight million. Many were executed, most died from malnourishment and exhaustion as a result of the famine caused by inept and negligent government officials.

China

Chinese scholars had kept count of 1,828 instances of famine from 108 BC to 1911 in one province or another — an average of close to one famine per year. From 1333 to 1337 a terrible famine killed 6 million Chinese. The four famines of 1810, 1811, 1846, and 1849 are said to have killed no fewer than 45 million people. The period from 1850 to 1873 saw, as a result of the Taiping Rebellion, drought, and famine, the population of China drop by over 60 million people. China's Qing Dynasty bureaucracy, which devoted extensive attention to minimizing famines, is credited with averting a series of famines following El Niño-Southern Oscillation-linked droughts and floods. These events are comparable, though somewhat smaller in scale, to the ecological trigger events of China's vast 19th-century famines. (Pierre-Etienne Will, *Bureaucracy and Famine*) Qing China carried out its relief efforts, which included vast shipments of food, a requirement that the rich open their storehouses to the poor, and price regulation, as part of a state guarantee of subsistence to the peasantry (known as *ming-sheng*).

When a stressed monarchy shifted from state management and direct shipments of grain to monetary charity in the mid-19th century, the system broke down. Thus the 1867–68 famine under the Tongzhi Restoration was successfully relieved but the Great North China Famine of 1877–78, caused by drought across northern China, was a catastrophe. The province of Shanxi was substantially depopulated as grains ran out, and desperately starving people stripped forests, fields, and their very houses for food. Estimated mortality is 9.5 to 13 million people. (Mike Davis, *Late Victorian Holocausts*)

Great Leap Forward

The largest famine of the 20th century, and almost certainly of all time, was the 1958–61 Great Leap Forward famine in China. The immediate causes of this famine lay in Mao Zedong's ill-fated attempt to transform China from an agricultural nation to an industrial power in one huge leap. Communist Party cadres across China insisted that peasants abandon their farms for collective farms, and begin to produce steel in small foundries, often melting down their farm instruments in the process.

Collectivisation undermined incentives for the investment of labour and resources in agriculture; unrealistic plans for decentralized metal production sapped needed labour; unfavourable weather conditions; and communal dining halls encouraged overconsumption of available food, "Communal dining and the Chinese Famine 1958-1961"). Such was the centralized control of information and the intense pressure on party cadres to report only good news—such as production quotas met or exceeded—that information about the escalating disaster was effectively suppressed. When the leadership did become aware of the scale of the famine, it did little to respond, and continued to ban any discussion of the cataclysm. This blanket suppression of news was so effective that very few Chinese citizens were aware of the scale of the famine, and the greatest peacetime demographic disaster of the 20th century only became widely known twenty years later, when the veil of censorship began to lift.

The exact number of famine deaths during 1958–61 is difficult to determine, and estimates range from 18 to at least 42 million people, with a further 30 million cancelled or delayed births. It was only when the famine had wrought its worst that Mao was forced to reverse agricultural collectivisation policies, which were effectively dismantled in 1978. China has not experienced a famine of the proportions of the Great Leap Forward since 1961.

India

Owing to its almost entire dependence upon the monsoon rains, India is vulnerable to crop failures, which upon occasion deepen into famine. There were 14 famines in India between the 11th and 17th centuries (Bhatia, 1985). For example, during the 1022–1033 Great famines in India entire provinces were depopulated. Famine in Deccan killed at least 2 million people in 1702-1704. B.M. Bhatia believes that the earlier famines were localised, and it was only after 1860, during the British rule, that famine came to signify general shortage of foodgrains in the country. There were approximately 25 major famines spread through states such as Tamil Nadu in the south, and Bihar and Bengal in the east during the latter half of the 19th century.

Romesh Chunder Dutt argued as early as 1900, and present-day scholars such as Amartya Sen agree, that some historic famines were a product of both uneven rainfall and British economic and administrative policies, which since 1857 had led to the seizure and conversion of local farmland to foreign-owned plantations, restrictions on internal trade, heavy taxation of Indian citizens to support British expeditions in Afghanistan, inflationary measures that increased the price of food, and substantial exports of staple crops from India to Britain. (Dutt, 1900 and 1902; Srivastava, 1968; Sen, 1982; Bhatia, 1985.) Some British citizens, such as William Digby, agitated for policy reforms and famine relief, but Lord Lytton, the governing British viceroy in India, opposed such changes in the belief that they would stimulate shirking by Indian workers.

The first, the Bengal famine of 1770, is estimated to have taken around 10 million lives — one-third of Bengal's population at the time. Other notable famines include the Great Famine of 1876–78, in which 6.1 million to 10.3 million people died and the Indian famine of 1899–1900, in which 1.25 to 10 million people died. The famines continued until independence in 1947, with the Bengal Famine of 1943–44— even though there were no crop failures —killing 1.5 million to 3 million Bengalis during World War II.

The observations of the Famine Commission of 1880 support the notion that food distribution is more to blame for famines than food scarcity. They observed that each province in British India, including Burma, had a surplus of foodgrains, and the annual surplus was 5.16 million tons (Bhatia, 1970). At that time, annual export of rice and other grains from India was approximately one million tons.

"Population growth worsened the plight of the peasantry. As a result of peace and improved sanitation and health, the Indian population rose from perhaps 100 million in 1700 to 300 million by 1920. While encouraging agricultural productivity, the British also provided economic incentives to have more children to help in the fields. Although a similar population increase occurred in Europe at the same time, the growing numbers could be absorbed by industrialization or emigration to the Americas and Australia. India enjoyed neither an industrial revolution nor an increase in food growing. Moreover, Indian landlords had a stake in the cash crop system and discouraged innovation. As a result, population numbers far outstripped the amount of available food and land, creating dire poverty and widespread hunger . "

—-Craig A. Lockard, *Societies, Networks, and Transitions*

In 1966, there was a close call in Bihar, when the United States allocated 900,000 tons of grain to fight the famine. Three years of drought in India resulted in an estimated 1.5 million deaths from starvation and disease.

4

The Socio-Economic Impacts of Climate Change on Agriculture

Climate change impacts can be measured as an economic cost (Smith *et al.*, 2001:936-941). This is particularly well-suited to market impacts, that is impacts that are linked to market transactions and directly affect GDP. Monetary measures of non-market impacts, e.g., impacts on human health and ecosystems, are more difficult to calculate. Other difficulties with impact estimates are listed below:

- Knowledge gaps: Calculating distributional impacts requires detailed geographical knowledge, but these are a major source of uncertainty in climate models.
- Vulnerability: Compared with developed countries, there is a limited understanding of the potential market sector impacts of climate change in developing countries.
- Adaptation: The future level of adaptive capacity in human and natural systems to climate change will affect how society will be impacted by climate change. Assessments may under- or overestimate adaptive capacity, leading to under- or overestimates of positive or negative impacts.
- Socioeconomic trends: Future predictions of development affect estimates of future climate change impacts, and in some instances, different estimates of development trends lead to a reversal from a predicted positive, to a predicted negative, impact (and *vice versa*).

In a literature assessment, Smith *et al.* (2001:957-958) concluded, with medium confidence, that:

- climate change would increase income inequalities between and within countries.
- a small increase in global mean temperature (up to 2 °C, measured against 1990 levels) would result in net negative market sector impacts in many developing countries and net positive market sector impacts in many developed countries.

With high confidence, it was predicted that with a medium (2-3 °C) to high level of warming (greater than 3 °C), negative impacts would be exacerbated, and net positive impacts would start to decline and eventually turn negative.

Non-market Impacts

Smith *et al.* (2001:942) predicted that climate change would likely result in pronounced non-market impacts. Most of impacts were predicted to be negative. The literature assessmented by Smith *et al.* (2001) suggested that climate change would cause substantial negative health impacts in developing countries. Smith *et al.* (2001) noted that few of the studies they reviewed had adequately accounted for adaptation. In a literature assessment, Confalonieri *et al.* (2007:415) found that in the studies that had included health impacts, those impacts contributed substantially to the total costs of climate change.

Market Sector

Agriculture: Depending on underlying assumptions, studies of the economic impacts of a doubling in atmospheric carbon dioxide (CO_2) from pre-industrial levels conclude that this would have a slightly negative to moderately positive aggregate effect (i.e., total impacts across all regions) on the agricultural sector (Smith *et al.*, 2001:938). This aggregate effect hides substantial regional differences, with benefits mostly predicted in the developed world and strongly negative impacts for populations poorly connected to regional and global trading systems.

Fishery

The fishing industry has been affected by climate change. Evidence of past fish abundance and sizes can be seen in old photographs. Loren McClenachan, a graduate student at the Scripps Institute of Oceanography in the University of California-San Diego, found a historical archive of old photographs, from Florida's Key West, at the Monroe Country Public Library. The archive contained pictures aboard a series of vessels called the *Gulf Stream and Greyhound.* Old photographs of catches showed bigger fishes and larger catches while recent photos have shown decreased fish abundance and size. Research suggests that recreational and commercial fishers may have led to chronic overfishing but historical records from other parts of the globe have shown a decline in their fish stocks as well.

Data from the FAO since 1950 revealed an increased in the world's total marine fish production. The graph began to level off around the 1990s. In 1995, a fishing fleet operational cost was $124 billion but revenues were only $70 billion. To make up for the shortfall, many governments had subsidized fishermen with an estimated $30–$34 billion a year to make up their shortfalls. The subsidy sent the wrong message to producers, boasting an unsustainable industry and leading to a misallocation of resources. The number of decked fishing vessels went up from around 600,000 in 1970 to 1,260,000 in 1995. If subsidies were removed, fishers would have to cut down on the number of fishing fleet to cut cost, thus avoiding overfishing and the possible collapse of the fishery.

Other Sectors

A number of other sectors will be affected by climate change, including the livestock, forestry, and fisheries industries. Other sectors sensitive to climate change include the energy, construction, insurance, tourism and recreation industries. The aggregate impact of climate change on most of these sectors is highly uncertain (Schneider *et al.*, 2007:790).

Regions

- Africa: In Africa, coastal facilities are economically significant. In a literature assessment, Desanker *et al.* (2001:490) concluded that climate change would result in sea-level rise, coastal erosion, saltwater intrusion, and flooding. Desanker *et al.* (2001) predicted that these changes would have a significant impact on African communities and economies.
- Coasts and low-lying areas: In literature assessment, Nicholls *et al.* (2007:338-339) concluded that the socio-economic impacts of climate change on coastal and low-lying areas would be overwhelmingly adverse. Some benefits, however, were noted, e.g., the opening of new ocean routes due to reduced sea ice. Compared with developed countries, the protection costs

associated with projected sea level rise were found to be relatively higher for developing countries.

- Polar regions: Anisimov *et al.* (2001:804) reviewed the literature on climate change impacts in polar regions. With very high confidence, they concluded that the impact of climate change on infrastructure would increase economic costs. New opportunities for trading and shipping across the Arctic ocean, lower operational costs for the oil and gas industry, lower heating costs, and easier access for ship-based tourism, were expected to bring economic benefits.
- Small islands: In a literature assessment, Mimura *et al.* (2007:689) concluded, with high confidence, that on small islands, tourism would, for the most part, be negatively affected by climate change. On many small islands, tourism is a major contributor to GDP and employment.

Other Systems and Sectors

- Freshwater resources: In this sector, costs and benefits of climate change may take several forms, including monetary costs and benefits, and ecosystem and human impacts, e.g., loss of aquatic species and household flooding. In a literature assessment, Kundzewicz *et al.* (2007:191) found that few of these costs had been estimated in monetary terms. In respect to the water supply, they predicted that costs would very likely exceed benefits. Predicted costs included the potential need for infrastructure investments to protect against floods and droughts.
- Industry, settlements and society:
  - In a literature assessment, Wilbanks *et al.* (2007:377) concluded, with high confidence, that the economic costs of extreme weather events, at large national or large regional scale, would be unlikely to exceed more than a few percent of the total economy in the year of the event, except for possible abrupt changes. In smaller locations, particularly developing countries, it was estimated with high confidence that, in the year of the extreme event, short-run damages could amount to more than 25% GDP.
  - Infrastructure: According to Tol (2008), roads, airport runways, railway lines and pipelines, (including oil

pipelines, sewers, water mains etc.) may require increased maintenance and renewal as they become subject to greater temperature variation and are exposed to weather that they were not designed for.

Aggregate Impacts

Aggregating impacts adds up the total impact of climate change across sectors and/or regions (IPCC, 2007a:76). In producing aggregate impacts, there are a number of difficulties, such as predicting the ability of societies to adapt climate change, and estimating how future economic and social development will progress (Smith *et al.*, 2001:941). It is also necessary for the researcher to make subjective value judgements over the importance of impacts occurring in different economic sectors, in different regions, and at different times.

Smith *et al.* (2001) assessed the literature on the aggregate impacts of climate change. With medium confidence, they concluded that a small increase in global average temperature (up to 2 °C, measured against 1990 levels) would result in an aggregate market sector impact of plus or minus a few percent of world GDP. Smith *et al.* (2001) found that for a small to medium (2-3 °C) global average temperature increase, some studies predicted small net positive market impacts. Most studies they assessed predicted net damages beyond a medium temperature increase, with further damages for greater (more than 3 °C) temperature rises.

With low confidence, Smith *et al.* (2001) concluded that the non-market impacts of climate change would be negative. Smith *et al.* (2001:942) decided that studies might have understated the true costs of climate change, e.g., by not correctly estimating the impact of extreme weather events. It was thought possible that some of the positive impacts of climate change had been overlooked, and that adaptive capacity had possibly been underestimated.

Some of the studies assessed by Schneider *et al.* (2007:790) predicted that gross world product could increase for 1-3 °C warming (relative to temperatures over the 1990-2000 period), largely because of aggregate benefits in the agricultural sector. In the view of Schneider *et al.* (2007), these estimates carried low confidence. Stern (2007) assessed climate change impacts using the basic economics of risk premiums (Yohe *et al.*, 2007:821). He found that unmitigated climate change could result in a reduction in welfare equivalent to a persistent average fall in global per-capita consumption of at least 5%. The study by Stern (2007) has received both criticism and support from other economists. IPCC (2007a) concluded that "Aggregate estimates of costs mask significant differences in impacts across sectors, regions and populations and *very likely* underestimate damage costs because they cannot include many non-quantifiable impacts."

Professor Nick Bostrom comments that "Even the Stern Review on the Economics of Climate Change, a report prepared for the British Government which has been criticized by some as overly pessimistic, estimates that under the assumption of business-as-usual with regard to emissions, global warming will reduce welfare by an amount equivalent to a permanent reduction in per capita consumption of between 5 and 20%. In absolute terms, this would be a huge harm. Yet over the course of the twentieth century, world GDP grew by some 3,700%, and per capita world GDP rose by some 860%. It seems safe to say that (absent a radical overhaul of our best current scientific models of the Earth's climate system) whatever negative economic effects global warming will have, they will be completely swamped by other factors that will influence economic growth rates in this century."

Marginal Impacts

The social cost of carbon (SCC) is an aggregate measure of the impacts of climate change. It is defined as the incremental (or marginal) social cost of emitting one more tonne of carbon (as carbon dioxide) into the atmosphere at any point in time (Yohe *et al.*, 2007:821). Different GHGs have different social costs. For example, due to their greater physical capacity to trap infrared radiation, HFCs have a considerably higher social cost per tonne of emission than carbon dioxide. Another physical property that affects the social cost is the atmospheric lifetime of the GHG.

Estimates of the SCC are given in the carbon tax article. These estimates are highly uncertain and cover a wide range (Klein *et al.*, 2007:756). The discrepancies in estimates can be broken down into normative and empirical parameters (Fisher *et al.*, 2007:232). Key normative parameters include the aggregation of impacts across time and regions. The other parameters relate to the empirical validity of SCC estimates. This reflects the poor quality of data on which estimates are based, and the difficulty in predicting how society will react to future climate change. In a literature assessment, Klein *et al.* (2007:757) placed low confidence in SCC estimates.

Sensitivity Analysis

Sensitivity analysis allows assumptions to be changed in aggregate analysis to see what effect it has on results (Smith *et al.*, 2001:943):

- Shape of the damage function: This relates impacts to the change in atmospheric greenhouse gas (GHG) concentrations. There is little information on what the correct shape (e.g., linear or cubic) of this function is. Compared with a linear function, a cubic function shows relatively small damages for small increases in temperature, but more sharply increasing damages at greater temperatures.
- Rate of climate change: This is believed to be an important determinant of impacts, often because it affects the time available for adaptation.
- Discount rate and time horizon: Models used in aggregate studies suggest that the most severe impacts of climate change will occur in the future. Estimated impacts are therefore sensitive to the time horizon (how far a given study projects impacts into the future) and the discount rate (the value assigned to consumption in the future versus consumption today).
- Welfare criteria: Aggregate analysis is particularly sensitive to the weighting (i.e., relative importance) of impacts occurring in different regions and at different times. Studies by Fankhauser *et al.* (1997) and Azar (1999) found that greater concern over the distribution of impacts lead to more severe predictions of aggregate impacts.
- Uncertainty: Usually assessed through sensitivity analysis, but can also be viewed as a hedging problem. EMF (1997) found that deciding on how to hedge depends on society's aversion to climate change risks, and the potential costs of insuring against these risks.

Advantages and Disadvantages

There are a number of benefits of using aggregated assessments to measure climate change impacts (Smith *et al.*, 2001:954). They allow impacts to be directly compared between different regions and times. Impacts can be compared with other environmental problems and also with the costs of avoiding those impacts. A problem of aggregated analyses is that they often reduce different types of impacts into a small number of indicators. It can be argued that some impacts are not well-suited to this, e.g., the monetization of mortality and loss of species diversity. On the other hand, Pearce (2003:364) argued that where there are monetary costs of avoiding impacts, it is not possible to avoid monetary valuation of those impacts.

Climate Change and Agriculture

Climate change and agriculture are interrelated processes, both of which take place on a global scale. Global warming is projected to have significant impacts on conditions affecting agriculture, including temperature, carbon dioxide, glacial run-off, precipitation and the interaction of these elements. These conditions determine the carrying capacity of the biosphere to produce enough food for the human population and domesticated animals. The overall effect of climate change on agriculture will depend on the balance of these effects. Assessment of the effects of global climate changes on agriculture might help to properly anticipate and adapt farming to maximize agricultural production.

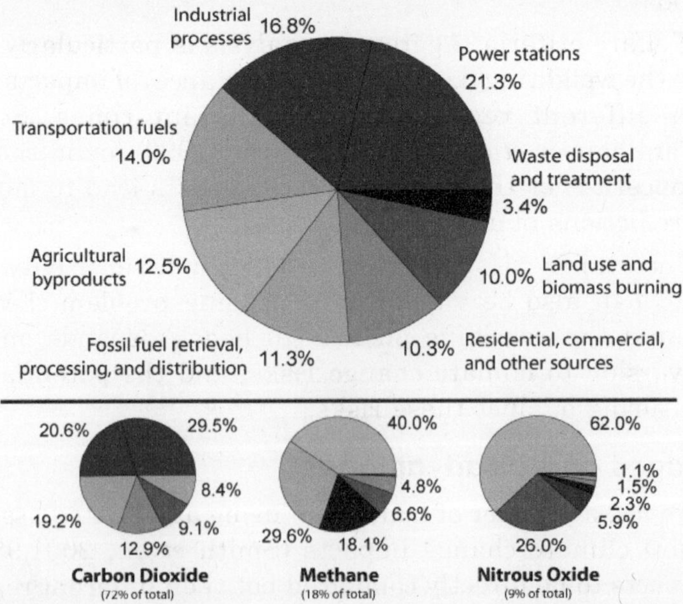

At the same time, agriculture has been shown to produce significant effects on climate change, primarily through the production and release of greenhouse gases such as carbon dioxide, methane, and nitrous oxide, but also by altering the Earth's land cover, which can change its ability to absorb or reflect heat and light, thus contributing to radiative forcing. Land use change such as deforestation and desertification, together with use of fossil fuels, are the major anthropogenic sources of carbon dioxide; agriculture itself is the major contributor to increasing methane and nitrous oxide concentrations in Earth's atmosphere.

Impact of Climate Change on Agriculture

Despite technological advances, such as improved varieties, genetically modified organisms, and irrigation systems, weather is still a key factor in agricultural productivity, as well as soil properties and natural communities. The effect of climate on agriculture is related to variabilities in local climates rather than in global climate patterns. The Earth's average surface temperature has increased by 1.5°F {0.83°C} since 1880. Consequently, agronomists consider any assessment has to be individually consider each local area.

On the other hand, agricultural trade has grown in recent years, and now provides significant amounts of food, on a national level to major importing countries, as well as comfortable income to exporting ones. The international aspect of trade and security in terms of food implies the need to also consider the effects of climate change on a global scale.

A study published in *Science* suggests that, due to climate change, "southern Africa could lose more than 30% of its main crop, maize, by 2030. In South Asia losses of many regional staples, such as rice, millet and maize could top 10%".

The Intergovernmental Panel on Climate Change (IPCC) has produced several reports that have assessed the scientific literature on climate change. The IPCC Third Assessment Report, published in 2001, concluded that the poorest countries would be hardest hit, with reductions in crop yields in most tropical and sub-tropical regions due to decreased water availability, and new or changed insect pest incidence. In Africa and Latin America many rainfed crops are near their maximum temperature tolerance, so that yields are likely to fall sharply for even small climate changes; falls in agricultural productivity of up to 30% over the 21st century are projected. Marine life and the fishing industry will also be severely affected in some places. Climate change induced by increasing greenhouse gases is likely to affect crops differently from region to region. For example, average crop yield is expected to drop down to 50% in Pakistan according to the UKMO scenario whereas corn production in Europe is expected to grow up to 25% in optimum hydrologic conditions.

More favourable effects on yield tend to depend to a large extent on realization of the potentially beneficial effects of carbon dioxide on crop growth and increase of efficiency in water use. Decrease in potential yields is likely to be caused by shortening of the growing period, decrease in water availability and poor vernalization.

In the long run, the climatic change could affect agriculture in several ways :

- *productivity*, in terms of quantity and quality of crops
- *agricultural practices*, through changes of water use (irrigation) and agricultural inputs such as herbicides, insecticides and fertilizers
- *environmental effects*, in particular in relation of frequency and intensity of soil drainage (leading to nitrogen leaching), soil erosion, reduction of crop diversity
- *rural space*, through the loss and gain of cultivated lands, land speculation, land renunciation, and hydraulic amenities.
- *adaptation*, organisms may become more or less competitive, as well as humans may develop urgency to develop more competitive organisms, such as flood resistant or salt resistant varieties of rice.

They are large uncertainties to uncover, particularly because there is lack of information on many specific local regions, and include the uncertainties on magnitude of climate change, the effects of technological changes on productivity, global food demands, and the numerous possibilities of adaptation.

Most agronomists believe that agricultural production will be mostly affected by the severity and pace of climate change, not so much by gradual trends in climate. If change is gradual, there may be enough time for biota adjustment. Rapid climate change, however, could harm agriculture in many countries, especially those that are already suffering from rather poor soil and climate conditions, because there is less time for optimum natural selection and adaption.

Observed Impacts

So far, the effects of regional climate change on agriculture have been relatively limited. Changes in crop phenology provide important evidence of the response to recent regional climate change. Phenology is the study of natural phenomena that recur periodically, and how these phenomena relate to climate and seasonal changes. A significant advance in phenology has been observed for agriculture and forestry in large parts of the Northern Hemisphere.

Droughts have been occurring more frequently because of global warming and they are expected to become more frequent and intense in Africa, southern Europe, the Middle East, most of the Americas, Australia, and Southeast Asia. Their impacts are aggravated because of increased water demand, population growth, urban expansion, and environmental protection efforts in many areas. Droughts result in crop failures and the loss of pasture grazing land for livestock.

Projections

As part of the IPCC's Fourth Assessment Report, Schneider *et al.* (2007) projected the potential future effects of climate change on agriculture. With low to medium confidence, they concluded that for about a 1 to 3 °C global mean temperature increase (by 2100, relative to the 1990–2000 average level) there would be productivity decreases for some cereals in low latitudes, and productivity increases in high latitudes. In the IPCC Fourth Assessment Report, "low confidence" means that a particular finding has about a 2 out of 10 chance of being correct, based on expert judgement. "Medium confidence" has about a 5 out of 10 chance of being correct. Over the same time period, with medium confidence, global production potential was projected to:

- increase up to around 3 °C,
- very likely decrease above about 3 °C.

Most of the studies on global agriculture assessed by Schneider *et al.* (2007) had not incorporated a number of critical factors, including changes in extreme events, or the spread of pests and diseases. Studies had also not considered the development of specific practices or technologies to aid adaptation to climate change.

Food Security

The IPCC Fourth Assessment Report also describes the impact of climate change on food security. Projections suggested that there could be large decreases in hunger globally by 2080, compared to the (then-current) 2006 level. Reductions in hunger were driven by projected social and economic development. For reference, the Food and Agriculture Organization has estimated that in 2006, the number of people undernourished globally was 820 million. Three scenarios *without* climate change (SRES A1, B1, B2) projected 100-130 million undernourished by the year 2080, while another scenario without climate change (SRES A2) projected 770 million undernourished. Based on an expert assessment of all of the evidence, these projections were thought to have about a 5-in-10 chance of being correct.

The same set of greenhouse gas and socio-economic scenarios were also used in projections that included the effects of climate change. *Including* climate change, three scenarios (SRES A1, B1, B2) projected 100-380 million undernourished by the year 2080, while another scenario with climate change (SRES A2) projected 740-1,300 million undernourished. These projections were thought to have between a 2-in-10 and 5-in-10 chance of being correct. Projections also suggested regional changes in the global distribution of hunger. By 2080, sub-Saharan Africa may overtake Asia as the world's most food-insecure region. This is mainly due to projected social and economic changes, rather than climate change.

"Climate change merely increases the urgency of reforming trade policies to ensure that global food security needs are met" said C. Bellmann, ICTSD Programmes Director. A 2009 ICTSD-IPC study by Jodie Keane suggests that climate change could cause farm output in sub-Saharan Africa to decrease by 12 percent by 2080 - although in some African countries this figure could be as much as 60 percent, with agricultural exports declining by up to one fifth in others. Adapting to climate change could cost the agriculture sector $14bn globally a year, the study finds.

Regional

Africa: In Africa, IPCC (2007:13) projected that climate variability and change would severely compromise agricultural production and access to food. This projection was assigned "high confidence."

Africa's geography makes it particularly vulnerable to climate change, and seventy per cent of the population rely on rain-fed agriculture for their livelihoods. Tanzania's official report on climate change suggests that the areas that usually get two rainfalls in the year will probably get more, and those that get only one rainy season will get far less. The net result is expected to be that 33% less maize—the country's staple crop—will be grown.

Asia

In East and Southeast Asia, IPCC (2007:13) projected that crop yields could increase up to 20% by the mid-21st century. In Central and South Asia, projections suggested that yields might decrease by up to 30%, over the same time period. These projections were assigned "medium confidence." Taken together, the risk of hunger was projected to remain very high in several developing countries.

More detailed analysis of rice yields by the International Rice Research Institute forecast 20% reduction in yields over the region per degree Celsius of temperature rise. Rice becomes sterile if exposed to temperatures above 35 degrees for more than one hour during flowering and consequently produces no grain.

Australia and New Zealand

Hennessy *et al.*. (2007:509) assessed the literature for Australia and New Zealand. They concluded that without further adaptation to climate change, projected impacts would likely be substantial: By 2030, production from agriculture and forestry was projected to decline over much of southern and eastern Australia, and over parts of eastern New Zealand; In New Zealand, initial benefits were projected close to major rivers and in western and southern areas. Hennessy *et al.*. (2007:509) placed high confidence in these projections.

Europe

With high confidence, IPCC (2007:14) projected that in Southern Europe, climate change would reduce crop productivity. In Central and Eastern Europe, forest productivity was expected to decline. In Northern Europe, the initial effect of climate change was projected to increase crop yields.

Latin America

With high confidence, IPCC (2007:14) projected that in drier areas of Latin America, productivity of some important crops would decrease and livestock productivity decline, with adverse consequences for food security. In temperate zones, soybean yields were projected to increase.

North America

A number of studies have been produced which assess the impacts of climate change on agriculture in North America. The IPCC Fourth Assessment Report of agricultural impacts in the region cites 26 different studies. With high confidence, IPCC (2007:14–15) projected that over the first few decades of this century, moderate climate change would increase aggregate yields of rain-fed agriculture by 5–20%, but with important variability among regions. Major challenges were projected for crops that are near the warm end of their suitable range or which depend on highly utilized water resources.

Droughts are becoming more frequent and intense in arid and semiarid western North America as temperatures have been rising, advancing the timing and magnitude of spring snow melt foods and reducing river flow volume in summer. Direct effects of climate change include increased heat and water stress, altered crop phenology, and disrupted symbiotic interactions. These effects may be exacerbated by climate changes in river flow, and the combined effects are likely to reduce the abundance of native trees in favour of non-native herbaceous and drought-tolerant competitors, reduce the habitat quality for many native animals, and slow litter decomposition and nutrient cycling. Climate change effects on human water demand and irrigation may intensify these effects.

United States

The US Global Change Research Programme (2009) assessed the literature on the impacts of climate change on agriculture in the United States:

- Many crops will benefit from increased atmospheric CO_2 concentrations and low levels of warming, but higher levels of warming will negatively affect growth and yields. Extreme events will likely reduce crop yields.
- Weeds. diseases and insect pests benefit from warming, and will require more attention in regards to pest and weed control.
- Increasing CO_2 concentrations will reduce the land's ability to supply adequate livestock feed. Increased heat, disease, and weather extremes will likely reduce livestock productivity.

According to a paper by Deschenes and Greenstone (2006), predicted increases in temperature and precipitation will have virtually no effect on the most important crops in the US.

Polar Regions (Arctic and Antarctic)

Anisimov *et al.*. (2007:655) assessed the literature for the polar region (Arctic and Antarctica). With medium confidence, they concluded that the benefits of a less severe climate were dependent on local conditions. One of these benefits was judged to be increased agricultural and forestry opportunities. For the *Guardian* newspaper, Brown (2005) reported on how climate change had affected agriculture in Iceland. Rising temperatures had made the widespread sowing of barley possible, which had been untenable twenty years ago. Some of the warming was due to a local (possibly temporary) effect via ocean currents from the Caribbean, which had also affected fish stocks.

Small Islands

In a literature assessment, Mimura *et al.* (2007:689) concluded that on small islands, subsistence and commercial agriculture would very likely be adversely affected by climate change. This projection was assigned "high confidence."

Shortage in Grain Production

Between 1996 and 2003, grain production has stabilized slightly over 1800 millions of tons. In 2000, 2001, 2002 and 2003, grain stocks have been dropping, resulting in a global grain harvest that was short of consumption by 93 millions of tons in 2003. By 2012, North American corn prices had risen to a record $8.34 per bushel in August, leaving 20 of the 211 U.S. ethanol fuel plants idle.

The Earth's average temperature has been rising since the late 1970s, with nine of the 10 warmest years on record occurring since 1995. In 2002, India and the United States suffered sharp harvest reductions because of record temperatures and drought. In 2003 Europe suffered very low rainfall throughout spring and summer, and a record level of heat damaged most crops from the United Kingdom and France in the Western Europe through Ukraine in the East. Bread prices have been rising in several countries in the region..

Poverty Impacts

Researchers at the Overseas Development Institute (ODI) have investigated the potential impacts climate change could have on agriculture, and how this would affect attempts at alleviating poverty in the developing world. They argued that the effects from moderate climate change are likely to be mixed for developing countries. However, the vulnerability of the poor in developing countries to short term impacts from climate change, notably the increased frequency and severity of adverse weather events is likely to have a negative impact. This, they say, should be taken into account when defining agricultural policy.

Mitigation and Adaptation in Developing Countries

The Intergovernmental Panel on Climate Change (IPCC) has reported that agriculture is responsible for over a quarter of total global greenhouse gas emissions. Given that agriculture's share in global gross domestic product (GDP) is about 4 percent, these figures suggest that agriculture is highly Green House Gas intensive. Innovative agricultural practices and technologies can play a role in climate mitigation and adaptation. This adaptation and mitigation potential is nowhere more pronounced than in developing countries where agricultural productivity remains low; poverty, vulnerability and food insecurity remain high; and the direct effects of climate change are expected to be especially harsh. Creating the necessary agricultural technologies and harnessing them to enable developing countries to adapt their agricultural systems to changing climate will require innovations in policy and institutions as well. In this context, institutions and policies are important at multiple scales.

Travis Lybbert and Daniel Sumner suggest six policy principles: (1) The best policy and institutional responses will enhance information flows, incentives and flexibility. (2) Policies and institutions that promote economic development and reduce poverty will often improve agricultural adaptation and may also pave the way for more effective climate change mitigation through agriculture. (3) Business as usual among the world's poor is not adequate. (4) Existing technology options must be made more available and accessible without overlooking complementary capacity and investments. (5) Adaptation and mitigation in agriculture will require local responses, but effective policy responses must also reflect global impacts and inter-linkages. (6) Trade will play a critical role in both mitigation and adaptation, but will itself be shaped importantly by climate change.

Crop Development Models

Models for climate behaviour are frequently inconclusive. In order to further study effects of global warming on agriculture, other types of models, such as *crop development models*, *yield prediction*, quantities of *water or fertilizer consumed*, can be used. Such models condense the knowledge accumulated of the climate, soil, and effects observed of the results of various agricultural practices. They thus could make it possible to test strategies of adaptation to modifications of the environment. Because these models are necessarily simplifying natural conditions (often based on the assumption that weeds, disease and insect pests are controlled), it is not clear whether the results they give will have an *in-field* reality. However, some results are partly validated with an increasing number of experimental results.

Other models, such as *insect and disease development* models based on climate projections are also used (for example simulation of aphid reproduction or septoria (cereal fungal disease) development).

Scenarios are used in order to estimate climate changes effects on crop development and yield. Each scenario is defined as a set of meteorological variables, based on generally accepted projections. For example, many models are running simulations based on doubled carbon dioxide projections, temperatures raise ranging from 1 °C up to 5 °C, and with rainfall levels an increase or decrease of 20%. Other parameters may include humidity, wind, and solar activity. Scenarios of crop models are testing farm-level adaptation, such as sowing date shift, climate adapted species (vernalisation need, heat and cold resistance), irrigation and fertilizer adaptation, resistance to disease. Most developed models are about wheat, maize, rice and soybean.

Temperature Potential Effect on Growing Period

Duration of crop growth cycles are above all, related to temperature. An increase in temperature will speed up development. In the case of an annual crop, the duration between sowing and harvesting will shorten (for example, the duration ın order to harvest corn could shorten between one and four weeks). The shortening of such a cycle could have an adverse effect on productivity because senescence would occur sooner.

Effect of Elevated Carbon Dioxide on Crops

Carbon dioxide is essential to plant growth. Rising CO_2 concentration in the atmosphere can have both positive and negative consequences. Increased CO_2 is expected to have positive physiological effects by increasing the rate of photosynthesis. Currently, the amount of carbon dioxide in the atmosphere is 380 parts per million. In comparison, the amount of oxygen is 210,000 ppm. This means that often plants may be starved of carbon dioxide as the enzyme that fixes CO_2, rubisco, also fixes oxygen in the process of photorespiration. The effects of an increase in carbon dioxide would be higher on C3 crops (such as wheat) than on C4 crops (such as maize), because the former is more susceptible to carbon dioxide shortage. Studies have shown that increased CO_2 leads to fewer stomata developing on plants which leads to reduced water usage. Under optimum conditions of temperature and humidity, the yield increase could reach 36%, if the levels of carbon dioxide are doubled.

Further, few studies have looked at the impact of elevated carbon dioxide concentrations on whole farming systems. Most models study the relationship between CO_2 and productivity in isolation from other factors associated with climate change, such as an increased frequency of extreme weather events, seasonal shifts, and so on.

In 2005, the Royal Society in London concluded that the purported benefits of elevated carbon dioxide concentrations are "likely to be far lower than previously estimated when factors such as increasing ground-level ozone are taken into account."

Effect on Quality

According to the IPCC's TAR, "The importance of climate change impacts on grain and forage quality emerges from new research. For rice, the amylose content of the grain—a major determinant of cooking quality—is increased under elevated CO_2" (Conroy et al., 1994). Cooked rice grain from plants grown in high-CO_2 environments would be firmer than that from today's plants. However, concentrations of iron and zinc, which are important for human nutrition, would be lower (Seneweera and Conroy, 1997). Moreover, the protein content of the grain decreases under combined increases of temperature and CO_2 (Ziska et al., 1997). Studies using FACE have shown that increases in CO_2 lead to decreased concentrations of micronutrients in crop plants. This may have knock-on effects on other parts of ecosystems as herbivores will need to eat more food to gain the same amount of protein. Studies have shown that higher CO_2 levels lead to reduced plant uptake of nitrogen (and a smaller number showing the same for trace elements such as zinc) resulting in crops with lower nutritional value. This would primarily impact on populations in poorer countries less able to compensate by eating more food, more varied diets, or possibly taking supplements.

Reduced nitrogen content in grazing plants has also been shown to reduce animal productivity in sheep, which depend on microbes in their gut to digest plants, which in turn depend on nitrogen intake.

Agricultural Surfaces and Climate Changes

Climate change may increase the amount of arable land in high-latitude region by reduction of the amount of frozen lands. A 2005 study reports that temperature in Siberia has increased three degree Celsius in average since 1960 (much more than the rest of the world). However, reports about the impact of global warming on Russian agriculture indicate conflicting probable effects : while they expect a northward extension of farmable lands, they also warn of possible productivity losses and increased risk of drought. Sea levels are expected to get up to one metre higher by 2100, though this projection is disputed. A rise in the sea level would result in an agricultural land loss, in particular in areas such as South East Asia. Erosion, submergence of shorelines, salinity of the water table due to the increased sea levels, could mainly affect agriculture through inundation of low-lying lands.

Low lying areas such as Bangladesh, India and Vietnam will experience major loss of rice crop if sea levels rise as expected by the end of the century. Vietnam for example relies heavily on its southern tip, where the Mekong Delta lies, for rice planting. Any rise in sea level of no more than a metre will drown several km^2 of rice paddies, rendering Vietnam incapable of producing its main staple and export of rice.

Erosion and Fertility

The warmer atmospheric temperatures observed over the past decades are expected to lead to a more vigorous hydrological cycle, including more extreme rainfall events. Erosion and soil degradation is more likely to occur. Soil fertility would also be affected by global warming. However, because the ratio of carbon to nitrogen is a constant, a doubling of carbon is likely to imply a higher storage of nitrogen in soils as nitrates, thus providing higher fertilizing elements for plants, providing better yields. The average needs for nitrogen could decrease, and give the opportunity of changing often costly fertilisation strategies.

Due to the extremes of climate that would result, the increase in precipitations would probably result in greater risks of erosion, whilst at the same time providing soil with better hydration, according to the intensity of the rain. The possible evolution of the organic matter in the soil is a highly contested issue: while the increase in the temperature would induce a greater rate in the production of minerals, lessening the soil organic matter content, the atmospheric CO_2 concentration would tend to increase it.

Potential Effects of Global Climate Change on Pests, Diseases and Weeds

A very important point to consider is that weeds would undergo the same acceleration of cycle as cultivated crops, and would also benefit from carbonaceous fertilization. Since most weeds are C3 plants, they are likely to compete even more than now against C4 crops such as corn. However, on the other hand, some results make it possible to think that weedkillers could gain in effectiveness with the temperature increase.

Global warming would cause an increase in rainfall in some areas, which would lead to an increase of atmospheric humidity and the duration of the wet seasons. Combined with higher temperatures, these could favour the development of fungal diseases. Similarly, because of higher temperatures and humidity, there could be an increased pressure from insects and disease vectors.

Glacier Retreat and Disappearance

The continued retreat of glaciers will have a number of different quantitative impacts. In areas that are heavily dependent on water runoff from glaciers that melt during the warmer summer months, a continuation of the current retreat will eventually deplete the glacial ice and substantially reduce or eliminate runoff. A reduction in runoff will affect the ability to irrigate crops and will reduce summer stream flows necessary to keep dams and reservoirs replenished.

Approximately 2.4 billion people live in the drainage basin of the Himalayan rivers. India, China, Pakistan, Afghanistan, Bangladesh, Nepal and Myanmar could experience floods followed by severe droughts in coming decades. In India alone, the Ganges provides water for drinking and farming for more than 500 million people. The west coast of North America, which gets much of its water from glaciers in mountain ranges such as the Rocky Mountains and Sierra Nevada, also would be affected.

Ozone and UV-B

Some scientists think agriculture could be affected by any decrease in stratospheric ozone, which could increase biologically dangerous ultraviolet radiation B. Excess ultraviolet radiation B can directly affect plant physiology and cause massive amounts of mutations, and indirectly through changed pollinator behaviour, though such changes are not simple to quantify. However, it has not yet been ascertained whether an increase in greenhouse gases would decrease stratospheric ozone levels. In addition, a possible effect of rising temperatures is significantly higher levels of ground-level ozone, which would substantially lower yields.

ENSO Effects on Agriculture

ENSO (El Niño Southern Oscillation) will affect monsoon patterns more intensely in the future as climate change warms up the ocean's water. Crops that lie on the equatorial belt or under the tropical Walker circulation, such as rice, will be affected by varying monsoon patterns and more unpredictable weather. Scheduled planting and harvesting based on weather patterns will become less effective.

Areas such as Indonesia where the main crop consists of rice will be more vulnerable to the increased intensity of ENSO effects in the future of climate change. University of Washington professor, David Battisti, researched the effects of future ENSO patterns on the Indonesian rice agriculture using [IPCC]'s 2007 annual report and 20 different logistical models mapping out climate factors such as wind pressure, sea-level, and humidity, and found that rice harvest will experience a decrease in yield. Bali and Java, which holds 55% of the rice yields in Indonesia, will be likely to experience 9–10% probably of delayed monsoon patterns, which prolongs the hungry season. Normal planting of rice crops begin in October and harevest by January. However, as climate change affects ENSO and consequently delays planting, harvesting will be late and in drier conditions, resulting in less potential yields.

Impact of Agriculture on Climate Change

The agricultural sector is a driving force in the gas emissions and land use effects thought to cause climate change. In addition to being a significant user of land and consumer of fossil fuel, agriculture contributes directly to greenhouse gas emissions through practices such as rice production and the raising of livestock; according to the Intergovernmental Panel on Climate Change, the three main causes of the increase in greenhouse gases observed over the past 250 years have been fossil fuels, land use, and agriculture.

There is significant concern about the impacts of climate change and its variability on agricultural production worldwide. First, issues of food security figure prominently in the list of human activities and ecosystem services under threat of dangerous anthropogenic interference on Earth's climate. Second, each country is naturally concerned with potential damages and benefits that may arise over the coming decades from climate change impacts on its territory as well as globally, since these will affect domestic and international policies, trading patterns, resource use, regional planning and ultimately the welfare of its people.

Land Use

Agriculture contributes to greenhouse gas increases through land use in four main ways:

- CO_2 releases linked to deforestation
- Methane releases from rice cultivation
- Methane releases from enteric fermentation in cattle
- Nitrous oxide releases from fertilizer application

Together, these agricultural processes comprise 54% of methane emissions, roughly 80% of nitrous oxide emissions, and virtually all carbon dioxide emissions tied to land use. The planet's major changes to land cover since 1750 have resulted from deforestation in temperate regions: when forests and woodlands are cleared to make room for fields and pastures, the albedo of the affected area increases, which can result in either warming or cooling effects, depending on local conditions. Deforestation also affects regional carbon reuptake, which can result in increased concentrations of CO_2, the dominant greenhouse gas. Land-clearing methods such as slash and burn compound these effects by burning biomatter, which directly releases greenhouse gases and particulate matter such as soot into the air.

Livestock

Livestock and livestock-related activities such as deforestation and increasingly fuel-intensive farming practices are responsible for over 18% of human-made greenhouse gas emissions, including:

- 9% of global carbon dioxide emissions
- 35–40% of global methane emissions (chiefly due to enteric fermentation and manure)
- 64% of global nitrous oxide emissions (chiefly due to fertilizer use.)

Livestock activities also contribute disproportionately to land-use effects, since crops such as corn and alfalfa are cultivated in order to feed the animals. Worldwide, livestock production occupies 70% of all land used for agriculture, or 30% of the land surface of the Earth. Climate change impact studies have shown that the productivity of agricultural activities is highly sensitive to climate change. The effect of changes in climate on agricultural activities both physical and economic has been shown to be significant for low input farming systems, such as subsistence farming in developing countries in Sub-Saharan Africa that are located in marginal areas and have the least capacity to adapt to changing climatic conditions.

Because most developing countries are heavily dependent on agriculture, the effect of climate change on their productive croplands is likely to threaten economic development and the welfare of the population. In addition, developing countries in tropical regions usually have large areas of poor marginal soils that are unusable for agriculture, which makes them particularly vulnerable to potential damage from environmental changes. The importance of agriculture for the economies of most African countries and the farming sector's reliance on the quality of rains during the rainy season make Zimbabwe and other countries in the region particularly sensitive to climate change and food insecure.

The effect of climate change on agricultural systems can be seen in the interaction between changes in climate variables and the stresses that result from actions taken to increase agricultural production. Impacts on crop yields, agricultural productivity and food security vary depending on the types of agricultural practices and systems. There is growing evidence that further increases in global warming leading to changes in main climate variables – temperature, precipitation, sea level rise, atmospheric carbon dioxide content and incidence of extreme events – may significantly affect African agricultural production, with the result that the livelihoods of subsistence farmers and pastoral peoples, who make-up a large portion of rural populations in Sub-Saharan Africa, could be negatively affected. For instance, in areas where temperatures are already warm, such as Zimbabwe and most parts of sub-Saharan Africa, further increases in temperature may actually slow down rather than stimulate plant growth, culminating in a general decrease in expected yield for most of the current food crops.

The indirect effect of the increased temperatures is the anticipated reduction in effective rainfall even when current amounts of rainfall are sustained, culminating in greater incidence of crop failure. Empirical agronomic studies in Zimbabwe have revealed that climate change has a negative effect on the agricultural performance of major crops. For instance Muchena (1994) and Magadza (1994) showed that a 2°C rise in ambient temperature and a rise of mean temperature by 4°C significantly lowered yields. In another study, Makadho (1996) assessed the potential effects of climate change on corn, using a Global Circulation Model (GCM) and the dynamic crop growth model CERES-maize, and the results indicated that maize production was expected to significantly decrease by approximately 11–17%, under conditions of both irrigation and non-irrigation. A reduced crop growth period because of increases in temperature, particularly during the grain filling and ripening stages, has been found to be the main factor contributing to decreased yields.

Agricultural Industrialization

Industrial farming is a form of modern farming that refers to the industrialized production of livestock, poultry, fish, and crops. The methods of industrial agriculture are technoscientific, economic, and political. They include innovation in agricultural machinery and farming methods, genetic technology, techniques for achieving economies of scale in production, the creation of new markets for consumption, the application of patent protection to genetic information, and global trade. These methods are widespread in developed nations and increasingly prevalent worldwide. Most of the meat, dairy, eggs, fruits, and vegetables available in supermarkets are produced using these methods of industrial agriculture.

Historical Development and Future Prospects

The birth of industrial agriculture more or less coincides with that of the Industrial Revolution in general. The identification of nitrogen, potassium, and phosphorus (referred to by the acronym NPK) as critical factors in plant growth led to the manufacture of synthetic fertilizers, making possible more intensive types of agriculture.

The discovery of vitamins and their role in animal nutrition, in the first two decades of the 20th century, led to vitamin supplements, which in the 1920s allowed certain livestock to be raised indoors, reducing their exposure to adverse natural elements. The discovery of antibiotics and vaccines facilitated raising livestock in concentrated, controlled animal feed operations by reducing diseases caused by crowding. Chemicals developed for use in World War II gave rise to synthetic pesticides. Developments in shipping networks and technology have made long-distance distribution of agricultural produce feasible.

Agricultural production across the world doubled four times between 1820 and 1975 to feed a global population of one billion human beings in 1800 and 6.5 billion in 2002. During the same period, the number of people involved in farming dropped as the process became more automated. In the 1930s, 24 percent of the American population worked in agriculture compared to 1.5 percent in 2002; in 1940, each farm worker supplied 11 consumers, whereas in 2002, each worker supplied 90 consumers. The number of farms has also decreased, and their ownership is more concentrated. In the U.S., four companies kill 81 percent of cows, 73 percent of sheep, 57 percent of pigs, and produce 50 percent of chickens, cited as an example of "vertical integration" by the president of the U.S. National Farmers' Union. In 1967, there were one million pig farms in America; as of 2002, there were 114,000, with 80 million pigs (out of 95 million) killed each year on factory farms, according to the U.S. National Pork Producers Council. According to the Worldwatch Institute, 74 percent of the world's poultry, 43 percent of beef, and 68 percent of eggs are produced this way.

According to Denis Avery of the agribusiness funded Hudson Institute, Asia increased its consumption of pork by 18 million tons in the 1990s. As of 1997, the world had a stock of 900 million pigs, which Avery predicts will rise to 2.5 billion pigs by 2050. He told the College of Natural Resources at the University of California, Berkeley that three billion pigs will thereafter be needed annually to meet demand. He writes: "For the sake of the environment, we had better hope those hogs are raised in big, efficient confinement systems."

British Agricultural Revolution

The British agricultural revolution describes a period of agricultural development in Britain between the 16th century and the mid-19th century, which saw a massive increase in agricultural productivity and net output. This in turn supported unprecedented population growth, freeing up a significant percentage of the workforce, and thereby helped drive the Industrial Revolution. How this came about is not entirely clear. In recent decades, historians cited four key changes in agricultural practices, enclosure, mechanization, four-field crop rotation, and selective breeding, and gave credit to a relatively few individuals.

Challenges and Issues

The challenges and issues of industrial agriculture for global and local society, for the industrial agriculture sector, for the individual industrial agriculture farm, and for animal rights include the costs and benefits of both current practices and proposed changes to those practices. This is a continuation of thousands of years of the invention and use of technologies in feeding ever growing populations.

When hunter-gatherers with growing populations depleted the stocks of game and wild foods across the Near East, they were forced to introduce agriculture. But agriculture brought much longer hours of work and a less rich diet than hunter-gatherers enjoyed. Further population growth among shifting slash-and-burn farmers led to shorter fallow periods, falling yields and soil erosion. Plowing and fertilizers were introduced to deal with these problems - but once again involved longer hours of work and degradation of soil resources.

While the point of industrial agriculture is lower cost products to create greater productivity thus a higher standard of living as measured by available goods and services, industrial methods have side effects both good and bad. Further, industrial agriculture is not some single indivisible thing, but instead is composed of numerous separate elements, each of which can be modified, and in fact is modified in response to market conditions, government regulation, and scientific advances.

So the question then becomes for each specific element that goes into an industrial agriculture method or technique or process: What bad side effects are bad enough that the financial gain and good side effects are outweighed? Different interest groups not only reach different conclusions on this, but also recommend differing solutions, which then become factors in changing both market conditions and government regulations.

Society

The major challenges and issues faced by society concerning industrial agriculture include:

Maximizing the benefits:

- Cheap and plentiful food
- Convenience for the consumer
- The contribution to our economy on many levels, from growers to harvesters to processors to sellers

while minimizing the downsides:

- Environmental and social costs
- Damage to fisheries
- Cleanup of surface and groundwater polluted with animal waste
- Increased health risks from pesticides
- Increased ozone pollution via methane byproducts of animals
- Global warming from heavy use of fossil fuels

Liabilities

Economic: Economic liabilities for industrial agriculture include the dependence on finite non-renewable fossil fuel energy resources, as an input in farm mechanization (equipment, machinery), for food processing and transportation, and as an input in agricultural chemicals. A future increase in energy prices as projected by the International Energy Agency is therefore expected to result in increase in food prices; and there is therefore a need to 'de-couple' non-renewable energy usage from agricultural production. Other liabilities include peak phosphate as finite phosphate reserves are currently a key input into chemical fertilizer for industrial agriculture.

Environment

Industrial agriculture uses huge amounts of water, energy, and industrial chemicals; increasing pollution in the arable land, usable water and atmosphere. Herbicides, insecticides, fertilizers, and animal waste products are accumulating in ground and surface waters. "Many of the negative effects of industrial agriculture are remote from fields and farms. Nitrogen compounds from the Midwest, for example, travel down the Mississippi to degrade coastal fisheries in the Gulf of Mexico. But other adverse effects are showing up within agricultural production systems — for example, the rapidly developing resistance among pests is rendering our arsenal of herbicides and insecticides increasingly ineffective.". Chemicals used in industrial agriculture, as well as the practice of monoculture, have also been implicated in Colony Collapse Disorder which has led to a collapse in bee populations. Agricultural production is highly dependent on bee pollination to pollinate many varieties of plants, fruits and vegetables.

Social

A study done for the US. Office of Technology Assessment conducted by the UC Davis Macrosocial Accounting Project concluded that industrial agriculture is associated with substantial deterioration of human living conditions in nearby rural communities.

Future increase in food commodity prices, driven by the energy price rises under peak oil and dependency of industrial agriculture on fossil fuels is expected to lead to increase in food prices which has particular impacts on poor people. An example of this can be seen in the 2007-2008 world food price crisis. Food price increases have a disproportionate impact on the poor as they spend a large proportion of their income on food.

Animals

"Concentrated animal feeding operations" or "intensive livestock operations", can hold large numbers (some up to hundreds of thousands) of animals, often indoors. These animals are typically cows, hogs, turkeys, or chickens. The distinctive characteristics of such farms is the concentration of livestock in a given space. The aim of the operation is to produce as much meat, eggs, or milk at the lowest possible cost and with the greatest level of food safety.

Food and water is supplied in place, and artificial methods are often employed to maintain animal health and improve production, such as therapeutic use of antimicrobial agents, vitamin supplements and growth hormones. Growth hormones are not used in chicken meat production nor are they used in the European Union for any animal. In meat production, methods are also sometimes employed to control undesirable behaviours often related to stresses of being confined in restricted areas with other animals. More docile breeds are sought (with natural dominant behaviours bred out for example), physical restraints to stop interaction, such as individual cages for chickens, or animals physically modified, such as the de-beaking of chickens to reduce the harm of fighting. Weight gain is encouraged by the provision of plentiful supplies of food to animals breed for weight gain.

The designation "confined animal feeding operation" in the U.S. resulted from that country's 1972 Federal Clean Water Act, which was enacted to protect and restore lakes and rivers to a "fishable, swimmable" quality. The United States Environmental Protection Agency (EPA) identified certain animal feeding operations, along with many other types of industry, as point source polluters of groundwater. These operations were designated as CAFOs and subject to special anti-pollution regulation.

In 17 states in the U.S., isolated cases of groundwater contamination has been linked to CAFOs. For example, the ten million hogs in North Carolina generate 19 million tons of waste per year. The U.S. federal government acknowledges the waste disposal issue and requires that animal waste be stored in lagoons. These lagoons can be as large as 7.5 acres (30,000 m^2). Lagoons not protected with an impermeable liner can leak waste into groundwater under some conditions, as can runoff from manure spread back onto fields as fertilizer in the case of an unforeseen heavy rainfall. A lagoon that burst in 1995 released 25 million gallons of nitrous sludge in North Carolina's New River. The spill allegedly killed eight to ten million fish.

The large concentration of animals, animal waste, and dead animals in a small space poses ethical issues to some consumers. Animal rights and animal welfare activists have charged that intensive animal rearing is cruel to animals. As they become more common, so do concerns about air pollution and ground water contamination, and the effects on human health of the pollution and the use of antibiotics and growth hormones.

According to the U.S. Centres for Disease Control and Prevention (CDC), farms on which animals are intensively reared can cause adverse health reactions in farm workers. Workers may develop acute and chronic lung disease, musculoskeletal injuries, and may catch infections that transmit from animals to human beings. These type of transmissions, however, and extremely rare, as zoonotic diseases are uncommon.

Crops

The projects within the Green Revolution spread technologies that had already existed, but had not been widely used outside of industrialized nations. These technologies included pesticides, irrigation projects, and synthetic nitrogen fertilizer.

The novel technological development of the Green Revolution was the production of what some referred to as "miracle seeds." Scientists created strains of maize, wheat, and rice that are generally referred to as HYVs or "high-yielding varieties." HYVs have an increased nitrogen-absorbing potential compared to other varieties. Since cereals that absorbed extra nitrogen would typically lodge, or fall over before harvest, semi-dwarfing genes were bred into their genomes. Norin 10 wheat, a variety developed by Orville Vogel from Japanese dwarf wheat varieties, was instrumental in developing Green Revolution wheat cultivars. IR8, the first widely implemented HYV rice to be developed by the International Rice Research Institute, was created through a cross between an Indonesian variety named "Peta" and a Chinese variety named "Dee Geo Woo Gen." With the availability of molecular genetics in Arabidopsis and rice the mutant genes responsible (*reduced height(rht)*, *gibberellin insensitive (gai1)* and *slender rice (slr1)*) have been cloned and identified as cellular signalling components of gibberellic acid, a phytohormone involved in regulating stem growth via its effect on cell division. Stem growth in the mutant background is significantly reduced leading to the dwarf phenotype. Photosynthetic investment in the stem is reduced dramatically as the shorter plants are inherently more stable mechanically. Assimilates become redirected to grain production, amplifying in particular the effect of chemical fertilizers on commercial yield.

HYVs significantly outperform traditional varieties in the presence of adequate irrigation, pesticides, and fertilizers. In the absence of these inputs, traditional varieties may outperform HYVs. One criticism of HYVs is that they were developed as F1 hybrids, meaning they need to be purchased by a farmer every season rather than saved from previous seasons, thus increasing a farmer's cost of production.

Sustainable Agriculture

The idea and practice of sustainable agriculture has arisen in response to the problems of industrial agriculture. Sustainable agriculture integrates three main goals: environmental stewardship, farm profitability, and prosperous farming communities. These goals have been defined by a variety of disciplines and may be looked at from the vantage point of the farmer or the consumer.

Organic Farming Methods

Organic farming methods combine some aspects of scientific knowledge and highly limited modern technology with traditional farming practices; accepting some of the methods of industrial agriculture while rejecting others. Organic methods rely on naturally occurring biological processes, which often take place over extended periods of time, and a holistic approach; while chemical-based farming focuses on immediate, isolated effects and reductionist strategies.

Integrated Multi-Trophic Aquaculture is an example of this holistic approach. Integrated Multi-Trophic Aquaculture (IMTA) is a practice in which the by-products (wastes) from one species are recycled to become inputs (fertilizers, food) for another. Fed aquaculture (e.g. fish, shrimp) is combined with inorganic extractive (e.g. seaweed) and organic extractive (e.g. shellfish) aquaculture to create balanced systems for environmental sustainability (biomitigation), economic stability (product diversification and risk reduction) and social acceptability (better management practices).

Effects of Industrial Agriculture on Climate Change and the Mitigation Potential of Small-scale Agro-ecological Farms

According to the Intergovernmental Panel on Climate Change (IPCC), agriculture is responsible for 10-12% of total global anthropogenic emissions and almost a quarter of the continuing increase of greenhouse gas (GHG) emissions. Not all forms of agriculture, however, have equivalent impacts on global warming. Industrial agriculture contributes significantly to global warming, representing a large majority of total agriculture-related GHG emissions. Alternatively, ecologically based methods for agricultural production, predominantly used on small-scale farms, are far less energy-consumptive and release fewer GHGs than industrial agricultural production. Besides generating fewer direct emissions, agro-ecological management techniques have the potential to sequester more GHGs than industrial agriculture.

Here, we review the literature on the contributions of agriculture to climate change and show the extent of GHG contributions from the industrial agricultural system and the potential of agro-ecological smallholder agriculture to help reduce GHG emissions. These reductions are achieved in three broad areas when compared with the industrial agricultural system: (1) a decrease in materials used and fluxes involved in the release of GHGs based on agricultural crop management choices; (2) a decrease in fluxes involved in livestock production and pasture management; and (3) a reduction in the transportation of agricultural inputs, outputs and products through an increased emphasis on local food systems. Although there are a number of barriers and challenges towards adopting small-scale agro-ecological methods on the large scale, appropriate incentives can lead to incremental steps towards agro-ecological management that may be able to reduce and mitigate GHG emissions from the agricultural sector.

Industrial farming methods contribute to climate change. Whether you believe the evidence of global atmospheric warming or not, there is overwhelming evidence that changes are occurring in our climate. While climate change is a natural occurrence over geological time, we are currently seeing some rapid fluctuations in weather around the globe. We have been bombarded with messages about how we need to change our lifestyles in the developed countries to do our part to damp down these changes. There are three main gases that contribute to the Earth's greenhouse effect. Without the balanced interaction of these gases, Earth would be a much colder place and life as we know it would not exist. Carbon dioxide, methane and nitrous oxide help trap the sun's radiant heat inside our atmosphere, much like a greenhouse traps heat, or a closed car's interior on a summer's day warms up and holds heat in.

Most of us are well aware of the part that carbon dioxide plays in atmospheric warming. The burning of fossil fuels like coal since the Industrial Revolution has accelerated. Forests and humus in soils act like carbon banks to sequester some of this carbon. We have been made aware that our private vehicle use has added to the carbon load.

Industrial style farming adds all three greenhouse gases to the atmosphere. Tracts of forest cleared for huge plantations like the palm oil industry release carbon when burned. This combines with the oxygen in the air to form carbon dioxide. The industrial practices often deplete the humus storage system also releasing more carbon into the atmosphere. The new Green Revolution crops demand fertilizers to yield properly. Nitrogen fertilizers are liberally used to "push" the plants. Extra nitrogen not used by the plants combines with oxygen to form nitrous oxide which helps warm the planet. We have all laughed at the idea that cow farts are warming the planet. Actually cow farts are nearly pure methane, another greenhouse gas. The practise of feeding large grain rations to confined livestock produces methane in their digestive systems. Traditional farming methods do not confine livestock but allow it to forage and consume little or no grain and the animals fart far less.

What we can do Everybody, no matter where they live - city, suburb, rural - has a part to play in minimizing climate change. A relatively easy change is shopping at your local super market. Buy fewer processed foods with palm oil in them. Ask your grocer where the food was produced if they don't already label it and support buying local or at least food grown on your own continent. Refuse to buy animal products produced by industrial methods. Small changes in your food buying habits can add up to big changes in our atmosphere.

According to the National Academy of Sciences, the Earth's surface temperature has risen by about 1 degree Fahrenheit in the past century, with accelerated warming during the past two decades. There is new and stronger evidence that most of the warming over the last 50 years is attributable to human activities. Human activities have altered the chemical composition of the atmosphere through the buildup of greenhouse gases – primarily carbon dioxide, methane, and nitrous oxide. The heat-trapping property of these gases is undisputed although uncertainties exist about exactly how earth's climate responds to them.

Our Changing Atmosphere

Energy from the sun drives the earth's weather and climate, and heats the earth's surface; in turn, the earth radiates energy back into space. Atmospheric greenhouse gases (water vapour, carbon dioxide, and other gases) trap some of the outgoing energy, retaining heat somewhat like the glass panels of a greenhouse.

Without this natural "greenhouse effect," temperatures would be much lower than they are now, and life as known today would not be possible. Instead, thanks to greenhouse gases, the earth's average temperature is a more hospitable 60°F. However, problems may arise when the atmospheric concentration of greenhouse gases increases.

Since the beginning of the industrial revolution, atmospheric concentrations of carbon dioxide have increased nearly 30%, methane concentrations have more than doubled, and nitrous oxide concentrations have risen by about 15%. These increases have enhanced the heat-trapping capability of the earth's atmosphere. Sulphate aerosols, a common air pollutant, cool the atmosphere by reflecting light back into space; however, sulphates are short-lived in the atmosphere and vary regionally.

Why are greenhouse gas concentrations increasing? Scientists generally believe that the combustion of fossil fuels and other human activities are the primary reason for the increased concentration of carbon dioxide.

Plant respiration and the decomposition of organic matter release more than 10 times the CO_2 released by human activities; but these releases have generally been in balance during the centuries leading up to the industrial revolution with carbon dioxide absorbed by terrestrial vegetation and the oceans.

What has changed in the last few hundred years is the additional release of carbon dioxide by human activities. Fossil fuels burned to run cars and trucks, heat homes and businesses, and power factories are responsible for about 98% of U.S. carbon dioxide emissions, 24% of methane emissions, and 18% of nitrous oxide emissions. Increased agriculture, deforestation, landfills, industrial production, and mining also contribute a significant share of emissions. In 1997, the United States emitted about one-fifth of total global greenhouse gases.

Uncertainties

Like many fields of scientific study, there are uncertainties associated with the science of global warming. This does not imply that all things are equally uncertain. Some aspects of the science are based on well-known physical laws and documented trends, while other aspects range from 'near certainty' to 'big unknowns'.

What's Known for Certain?

Scientists know for certain that human activities are changing the composition of Earth's atmosphere. Increasing levels of greenhouse gases, like carbon dioxide (CO_2), in the atmosphere since pre-industrial times have been well documented. There is no doubt this atmospheric buildup of carbon dioxide and other greenhouse gases is largely the result of human activities.

It's well accepted by scientists that greenhouse gases trap heat in the Earth's atmosphere and tend to warm the planet. By increasing the levels of greenhouse gases in the atmosphere, human activities are strengthening Earth's natural greenhouse effect. The key greenhouse gases emitted by human activities remain in the atmosphere for periods ranging from decades to centuries.

A warming trend of about 1°F has been recorded since the late 19th century. Warming has occurred in both the northern and southern hemispheres, and over the oceans. Confirmation of 20th-century global warming is further substantiated by melting glaciers, decreased snow cover in the northern hemisphere and even warming below ground.

What's Likely but not Certain?

Figuring out to what extent the human-induced accumulation of greenhouse gases since pre-industrial times is responsible for the global warming trend is not easy. This is because other factors, both natural and human, affect our planet's temperature. Scientific understanding of these other factors – most notably natural climatic variations, changes in the sun's energy, and the cooling effects of pollutant aerosols – remains incomplete.

Nevertheless, the Intergovernmental Panel on Climate Change (IPCC) stated there was a "discernible" human influence on climate; and that the observed warming trend is "unlikely to be entirely natural in origin." In the most recent Third Assessment Report (2001), IPCC wrote "There is new and stronger evidence that most of the warming observed over the last 50 years is attributable to human activities."

In short, scientists think rising levels of greenhouse gases in the atmosphere are contributing to global warming, as would be expected; but to what extent is difficult to determine at the present time.

As atmospheric levels of greenhouse gases continue to rise, scientists estimate average global temperatures will continue to rise as a result. By how much and how fast remain uncertain. IPCC projects further global warming of 2.2-10°F (1.4-5.8°C) by the year 2100. This range results from uncertainties in greenhouse gas emissions, the possible cooling effects of atmospheric particles such as sulphates, and the climate's response to changes in the atmosphere.

The IPCC states that even the low end of this warming projection "would probably be greater than any seen in the last 10,000 years, but the actual annual to decadal changes would include considerable natural variability."

What are the Big Unknowns?

Scientists have identified that our health, agriculture, water resources, forests, wildlife and coastal areas are vulnerable to the changes that global warming may bring. But projecting what the exact impacts will be over the 21st century remains very difficult. This is especially true when one asks how a local region will be affected.

Scientists are more confident about their projections for large-scale areas (e.g., global temperature and precipitation change, average sea level rise) and less confident about the ones for small-scale areas (e.g., local temperature and precipitation changes, altered weather patterns, soil moisture changes). This is largely because the computer models used to forecast global climate change are still ill-equipped to simulate how things may change at smaller scales.

Some of the largest uncertainties are associated with events that pose the greatest risk to human societies. IPCC cautions, "Complex systems, such as the climate system, can respond in non-linear ways and produce surprises." There is the possibility that a warmer world could lead to more frequent and intense storms, including hurricanes. Preliminary evidence suggests that, once hurricanes do form, they will be stronger if the oceans are warmer due to global warming. However, the jury is still out whether or not hurricanes and other storms will become more frequent.

More and more attention is being aimed at the possible link between El Niño events – the periodic warming of the equatorial Pacific Ocean – and global warming. Scientists are concerned that the accumulation of greenhouse gases could inject enough heat into Pacific waters such that El Niño events become more frequent and fierce. Here too, research has not advanced far enough to provide conclusive statements about how global warming will affect El Niño.

Living with Uncertainty

Like many pioneer fields of research, the current state of global warming science can't always provide definitive answers to our questions. There is certainty that human activities are rapidly adding greenhouse gases to the atmosphere, and that these gases tend to warm our planet. This is the basis for concern about global warming.

The fundamental scientific uncertainties are these: How much more warming will occur? How fast will this warming occur? And what are the potential adverse and beneficial effects? These uncertainties will be with us for some time, perhaps decades. Global warming poses real risks. The exact nature of these risks remains uncertain. Ultimately, this is why we have to use our best judgement – guided by the current state of science – to determine what the most appropriate response to global warming should be.

Greenhouse Gasses

Estimating future emissions is difficult, because it depends on demographic, economic, technological, policy, and institutional developments. Several emissions scenarios have been developed based on differing projections of these underlying factors. For example, by 2100, in the absence of emissions control policies, carbon dioxide concentrations are projected to be 30-150% higher than today's levels.

Changing Climate

Global mean surface temperatures have increased 0.5-1.0°F since the late 19th century. The 20th century's 10 warmest years all occurred in the last 15 years of the century. Of these, 1998 was the warmest year on record. The snow cover in the Northern Hemisphere and floating ice in the Arctic Ocean have decreased. Globally, sea level has risen 4-8 inches over the past century. Worldwide precipitation over land has increased by about one percent. The frequency of extreme rainfall events has increased throughout much of the United States.

Increasing concentrations of greenhouse gases are likely to accelerate the rate of climate change. Scientists expect that the average global surface temperature could rise 1-4.5°F (0.6-2.5°C) in the next fifty years, and 2.2-10°F (1.4-5.8°C) in the next century, with significant regional variation. Evaporation will increase as the climate warms, which will increase average global precipitation. Soil moisture is likely to decline in many regions, and intense rainstorms are likely to become more frequent. Sea level is likely to rise two feet along most of the U.S. coast.

Calculations of climate change for specific areas are much less reliable than global ones, and it is unclear whether regional climate will become more variable. Once, all climate changes occurred naturally. However, during the Industrial Revolution, we began altering our climate and environment through changing agricultural and industrial practices. Before the Industrial Revolution, human activity released very few gases into the atmosphere, but now through population growth, fossil fuel burning, and deforestation, we are affecting the mixture of gases in the atmosphere.

What Are Greenhouse Gases?

Some greenhouse gases occur naturally in the atmosphere, while others result from human activities. Naturally occuring greenhouse gases include water vapour, carbon dioxide, methane, nitrous oxide, and ozone. Certain human activities, however, add to the levels of most of these naturally occurring gases:

- Carbon dioxide is released to the atmosphere when solid waste, fossil fuels (oil, natural gas, and coal), and wood and wood products are burned.
- Methane is emitted during the production and transport of coal, natural gas, and oil. Methane emissions also result from the decomposition of organic wastes in municipal solid waste landfills, and the raising of livestock.
- Nitrous oxide is emitted during agricultural and industrial activities, as well as during combustion of solid waste and fossil fuels.
- Very powerful greenhouse gases that are not naturally occurring include hydrofluorocarbons (HFCs), perfluorocarbons (PFCs), and sulfur hexafluoride (SF6), which are generated in a variety of industrial processes.

***Table ES-6**: Global Warming Potentials (100 Year Time Horizon)*

Gas	*GWP*
Carbon dioxide (CO2)	1
Methanc (CH4)*	21
Nitrous oxide (N2O)	310
HFC-23	11,700
HFC-125	2,800
HFC-134a	1,300
HFC-143a	3,800
HFC-152a	140
HFC-227ea	2,900
HFC-236fa	6,300
HFC-4310mee	1,300
CF4	6,500
C2F6	9,200
C4F10	7,000
C6F14	7,400
SF6	23,900

Each greenhouse gas differs in its ability to absorb heat in the atmosphere. HFCs and PFCs are the most heat-absorbent. Methane traps over 21 times more heat per molecule than carbon dioxide, and nitrous oxide absorbs 270 times more heat per molecule than carbon dioxide. Often, estimates of greenhouse gas emissions are presented in units of millions of metric tons of carbon equivalents (MMTCE), which weights each gas by its GWP value, or Global Warming Potential.

The concept of a Global Warming Potential (GWP) has been developed to compare the ability of each greenhouse gas to trap heat in the atmosphere relative to another gas. Carbon dioxide (CO2) was chosen as the reference gas to be consistent with IPCC guidelines.

Impacts

Rising global temperatures are expected to raise sea level, and change precipitation and other local climate conditions. Changing regional climate could alter forests, crop yields, and water supplies. It could also affect human health, animals, and many types of ecosystems. Deserts may expand into existing rangelands, and features of some of our National Parks may be permanently altered.

Most of the United States is expected to warm, although sulphates may limit warming in some areas. Scientists currently are unable to determine which parts of the United States will become wetter or drier, but there is likely to be an overall trend towards increased precipitation and evaporation, more intense rainstorms, and drier soils. Unfortunately, many of the potentially most important impacts depend upon whether rainfall increases or decreases, which can not be reliably projected for specific areas.

Actions

Today, action is occurring at every level to reduce, to avoid, and to better understand the risks associated with climate change. Many cities and states across the country have prepared greenhouse gas inventories; and many are actively pursuing programmes and policies that will result in greenhouse gas emission reductions.

5

Impacts of Sea Level Rise

Mean sea level (MSL) is a measure of the average height of the ocean's surface (such as the halfway point between the mean high tide and the mean low tide); used as a standard in reckoning land elevation. MSL also plays an important role in aviation, where standard sea level pressure is used as the measurement datum of altitude at flight levels.

Measurement

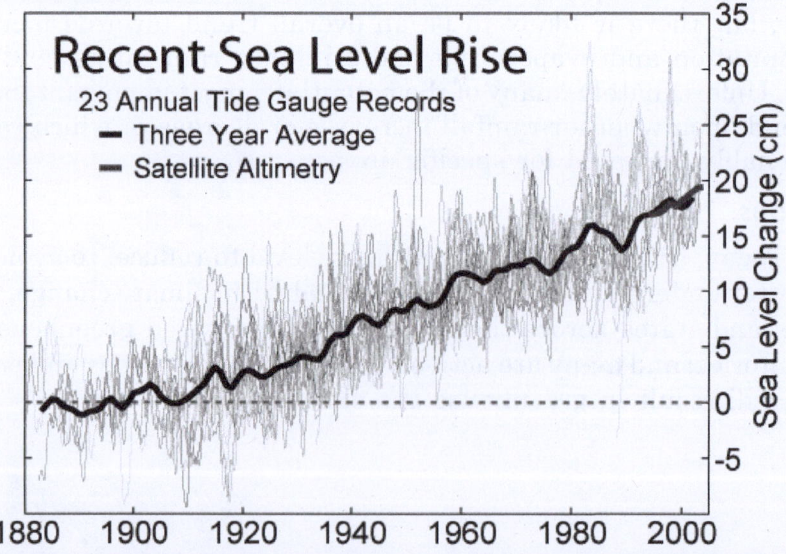

***Figure:** Sea level measurements from 23 long tide gauge records in geologically stable environments show a rise of around 200 millimetres (7.9 in) during the 20th century (2 mm/year).*

To an operator of a tide gauge, MSL means the "still water level"—the level of the sea with motions such as wind waves averaged out—averaged over a period of time such that changes in sea level, e.g., due to the tides, also get averaged out. One measures the values of MSL in respect to the land. Hence a change in MSL can result from a real change in sea level, or from a change in the height of the land on which the tide gauge operates.

In the UK, the Ordnance Datum (the 0 metres height on UK maps) is the mean sea level measured at Newlyn in Cornwall between 1915 and 1921. Prior to 1921 the datum was MSL at the Victoria Dock, Liverpool.

In France, the Marégraphe in Marseilles measures continuously the sea level since 1883 and offers the longest collapsed data about the sea level. It is used for a part of continental Europe and main part of Africa as official sea level.

Satellite altimeters have been making precise measurements of sea level since the launch of TOPEX/Poseidon in 1992. A joint mission of NASA and CNES, TOPEX/Poseidon was followed by Jason-1 in 2001 and the Ocean Surface Topography Mission on the Jason-2 satellite in 2008.

Difficulties in Utilization

To extend this definition far from the sea means comparing the local height of the mean sea surface with a "level" reference surface, or datum, called the geoid. In a state of rest or absence of external forces, the mean sea level would coincide with this geoid surface, being an equipotential surface of the Earth's gravitational field.

In reality, due to currents, air pressure variations, temperature and salinity variations, etc., this does not occur, not even as a long term average. The location-dependent, but persistent in time, separation between mean sea level and the geoid is referred to as (stationary) ocean surface topography. It varies globally in a range of ± 2 m.

Traditionally, one had to process sea-level measurements to take into account the effect of the 228-month Metonic cycle and the 223-month eclipse cycle on the tides. Mean sea level is not constant over the surface of the Earth. For instance, mean sea level at the Pacific end of the Panama Canal stands 20 cm (7.9 in) higher than at the Atlantic end.

Sea Level and Dry Land

Several terms are used to describe the changing relationships between sea level and dry land. When the term "relative" is used, it means change relative to a fixed point in the sediment pile. The term "eustatic" refers to global changes in sea level relative to a fixed point, such as the centre of the earth, for example as a result of melting ice-caps. The term "steric" refers to global changes in sea level due to thermal expansion and salinity variations.

The term "isostatic" refers to changes in the level of the land relative to a fixed point in the earth, possibly due to thermal buoyancy or tectonic effects; it implies no change in the volume of water in the oceans. The melting of glaciers at the end of ice ages is one example of eustatic sea level rise. The subsidence of land due to the withdrawal of groundwater is an isostatic cause of relative sea level rise. Paleoclimatologists can track sea level by examining the rocks deposited along coasts that are very tectonically stable, like the east coast of North America. Areas like volcanic islands are experiencing relative sea level rise as a result of isostatic cooling of the rock which causes the land to sink.

On other planets that lack a liquid ocean, planetologists can calculate a "mean altitude" by averaging the heights of all points on the surface. This altitude, sometimes referred to as a "sea level", serves equivalently as a reference for the height of planetary features.

Sea Level Change

Local and Eustatic Sea Level

Local mean sea level (LMSL) is defined as the height of the sea with respect to a land benchmark, averaged over a period of time (such as a month or a year) long enough that fluctuations caused by waves and tides are smoothed out. One must adjust perceived changes in LMSL to account for vertical movements of the land, which can be of the same order (mm/yr) as sea level changes. Some land movements occur because of isostatic adjustment of the mantle to the melting of ice sheets at the end of the last ice age. The weight of the ice sheet depresses the underlying land, and when the ice melts away the land slowly rebounds. Changes in ground-based ice volume also affect local and regional sea levels by the readjustment of the geoid and true polar wander. Atmospheric pressure, ocean currents and local ocean temperature changes can affect LMSL as well.

Eustatic change (as opposed to local change) results in an alteration to the global sea levels due to changes in either the volume of water in the world oceans or net changes in the volume of the ocean basins.

Short Term and Periodic Changes

There are many factors which can produce short-term (a few minutes to 14 months) changes in sea level.

Periodic sea level changes		
Diurnal and semidiurnal astronomical tides	12–24 h P	0.2–10+ m
Long-period tides		
Rotational variations (Chandler wobble)	14 month P	
Meteorological and oceanographic fluctuations		
Atmospheric pressure	Hours to months	–0.7 to 1.3 m
Winds (storm surges)	1–5 days	Up to 5 m
Evaporation and precipitation (may also follow long-term pattern)	Days to weeks	
Ocean surface topography (changes in water density and currents)	Days to weeks	Up to 1 m
El Niño/southern oscillation	6 mo every 5–10 yr	Up to 0.6 m
Seasonal variations		
Seasonal water balance among oceans (Atlantic, Pacific, Indian)		
Seasonal variations in slope of water surface		
River runoff/floods	2 months	1 m
Seasonal water density changes (temperature and salinity)	6 months	0.2 m
Seiches		
Seiches (standing waves)	Minutes to hours	Up to 2 m
Earthquakes		
Tsunamis (generate catastrophic long-period waves)	Hours	Up to 10 m
Abrupt change in land level	Minutes	Up to 10 m

Long Term Changes

Various factors affect the volume or mass of the ocean, leading to long-term changes in eustatic sea level. The primary influence is that of temperature on seawater density and the amounts of water in rivers, lakes, glaciers, polar ice caps and sea ice. Over much longer geological timescales, changes in the shape of the oceanic basins and in land/sea distribution will also affect sea level.

Observational and modelling studies of mass loss from glaciers and ice caps indicate a contribution to sea-level rise of 0.2 to 0.4 mm/yr averaged over the 20th century. Over this last million years, whereas it was higher most of the time before then, sea level was lower than today. Sea level reached 120 metres below current sea level at the Last Glacial Maximum 19,000-20,000 years ago.

Glaciers and Ice Caps

Each year about 8 mm (0.3 inch) of water from the entire surface of the oceans falls onto the Antarctica and Greenland ice sheets as snowfall. If no ice returned to the oceans, sea level would drop 8 mm every year. To a first approximation, the same amount of water appeared to return to the ocean in icebergs and from ice melting at the edges. Scientists previously had estimated which is greater, ice going in or coming out, called the mass balance, important because it causes changes in global sea level. High-precision gravimetry from satellites in low-noise flight has since determined that in 2006, the Greenland and Antarctic ice sheets experienced a combined mass loss of 475 ± 158 Gt/yr, equivalent to 1.3 ± 0.4 mm/yr sea level rise. Notably, the acceleration in ice sheet loss from 1988-2006 was 21.9 ± 1 Gt/yr^2 for Greenland and 14.5 ± 2 Gt/yr^2 for Antarctica, for a combined total of 36.3 ± 2 Gt/yr^2. This acceleration is 3 times larger than for mountain glaciers and ice caps (12 ± 6 Gt/yr^2).

Ice shelves float on the surface of the sea and, if they melt, to first order they do not change sea level. Likewise, the melting of the northern polar ice cap which is composed of floating pack ice would not significantly contribute to rising sea levels. Because they are lower in salinity, however, their melting would cause a very small increase in sea levels, so small that it is generally neglected.

- Scientists previously lacked knowledge of changes in terrestrial storage of water. Surveying of water retention by soil absorption and by artificial reservoirs ("impoundment") show that a total of about 10,800 cubic kilometres (2,591 cubic miles) of water (just under the size of Lake Superior) has been impounded on land to date. Such impoundment masked about 30 mm (1.2 in) of sea level rise in that time.
- If small glaciers and polar ice caps on the margins of Greenland and the Antarctic Peninsula melt, the projected rise in sea level will be around 0.5 m (1 ft 7.7 in). Melting of the Greenland ice sheet would produce 7.2 m (23.6 ft) of sea-level rise, and melting of the Antarctic ice sheet would produce 61.1 m (200.5 ft) of sea level rise. The collapse of the grounded interior reservoir of the West Antarctic Ice Sheet would raise sea level by 5–6 m.
- The snowline altitude is the altitude of the lowest elevation interval in which minimum annual snow cover exceeds 50%. This ranges from about 5,500 metres (18,045 feet) above sea-level at the equator down to sea level at about 70° N&S latitude, depending on regional temperature amelioration effects. Permafrost then appears at sea level and extends deeper below sea level polewards.

- As most of the Greenland and Antarctic ice sheets lie above the snowline and/or base of the permafrost zone, they cannot melt in a timeframe much less than several millennia; therefore it is likely that they will not, through melting, contribute significantly to sea level rise in the coming century. They can, however, do so through acceleration in flow and enhanced iceberg calving.
- Climate changes during the 20th century are estimated from modelling studies to have led to contributions of between –0.2 and 0.0 mm/yr from Antarctica (the results of increasing precipitation) and 0.0 to 0.1 mm/yr from Greenland (from changes in both precipitation and runoff).
- Estimates suggest that Greenland and Antarctica have contributed 0.0 to 0.5 mm/yr over the 20th century as a result of long-term adjustment to the end of the last ice age.

The current rise in sea level observed from tide gauges, of about 1.8 mm/yr, is within the estimate range from the combination of factors above but active research continues in this field. The terrestrial storage term, thought to be highly uncertain, is no longer positive, and shown to be quite large.

Geological Influences

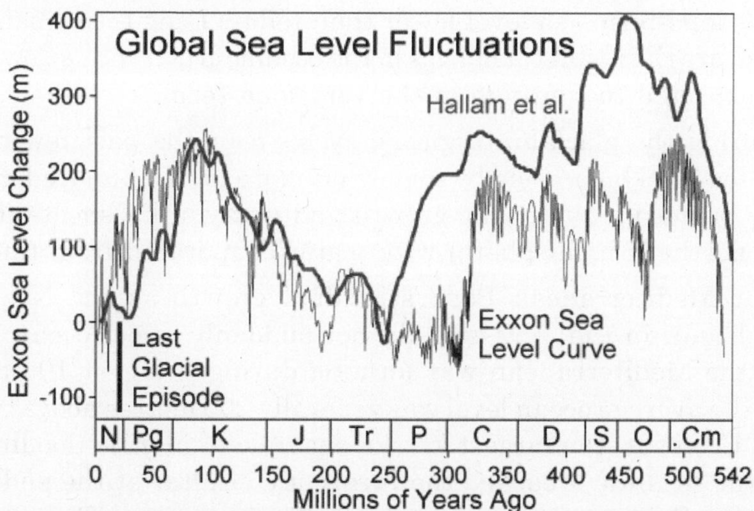

Figure: *Comparison of two sea level reconstructions during the last 500 Ma. The scale of change during the last glacial/interglacial transition is indicated with a black bar. Note that over most of geologic history, long-term average sea level has been significantly higher than today.*

At times during Earth's long history, the configuration of the continents and seafloor have changed due to plate tectonics.

This affects global sea level by determining the depths of the ocean basins and how glacial-interglacial cycles distribute ice across the Earth.

The depth of the ocean basins is a function of the age of oceanic lithosphere: as lithosphere becomes older, it becomes denser and sinks. Therefore, a configuration with many small oceanic plates that rapidly recycle lithosphere will produce shallower ocean basins and (all other things being equal) higher sea levels.

A configuration with fewer plates and more cold, dense oceanic lithosphere, on the other hand, will result in deeper ocean basins and lower sea levels.

When there were large amounts of continental crust near the poles, the rock record shows unusually low sea levels during ice ages, because there was lots of polar land mass upon which snow and ice could accumulate. During times when the land masses clustered around the equator, ice ages had much less effect on sea level.

Over most of geologic time, long-term sea level has been higher than today. Only at the Permian-Triassic boundary ~250 million years ago was long-term sea level lower than today. Long term changes in sea level are the result of changes in the oceanic crust, with a downward trend expected to continue in the very long term.

During the glacial/interglacial cycles over the past few million years, sea level has varied by somewhat more than a hundred metres. This is primarily due to the growth and decay of ice sheets (mostly in the northern hemisphere) with water evaporated from the sea.

The Mediterranean Basin's gradual growth as the Neotethys basin, begun in the Jurassic, did not suddenly affect ocean levels. While the Mediterranean was forming during the past 100 million years, the average ocean level was generally 200 metres above current levels. However, the largest known example of marine flooding was when the Atlantic breached the Strait of Gibraltar at the end of the Messinian Salinity Crisis about 5.2 million years ago. This restored Mediterranean sea levels at the sudden end of the period when that basin had dried up, apparently due to geologic forces in the area of the Strait.

Long-term causes	*Range of effect*	*Vertical effect*
Change in volume of ocean basins		
Plate tectonics and seafloor spreading (plate divergence/convergence) and change in seafloor elevation (mid-ocean volcanism)	Eustatic	0.01 mm/yr
Marine sedimentation	Eustatic	< 0.01 mm/yr
Change in mass of ocean water		
Melting or accumulation of continental ice	Eustatic	10 mm/yr
• Climate changes during the 20th century		
•• Antarctica	Eustatic	0.39 to 0.79 mm/yr[9]
•• Greenland (from changes in both precipitation and runoff)	Eustatic	0.0 to 0.1 mm/yr
• Long-term adjustment to the end of the last ice age		
•• Greenland and Antarctica contribution over 20th century	Eustatic	0.0 to 0.5 mm/yr
Release of water from earth's interior	Eustatic	
Release or accumulation of continental hydrologic reservoirs	Eustatic	
Uplift or subsidence of Earth's surface (Isostasy)		
Thermal-isostasy (temperature/density changes in earth's interior)	Local effect	
Glacio-isostasy (loading or unloading of ice)	Local effect	10 mm/yr
Hydro-isostasy (loading or unloading of water)	Local effect	
Volcano-isostasy (magmatic extrusions)	Local effect	
Sediment-isostasy (deposition and erosion of sediments)	Local effect	< 4 mm/yr
Tectonic uplift/subsidence		
Vertical and horizontal motions of crust (in response to fault motions)	Local effect	1–3 mm/yr
Sediment compaction		
Sediment compression into denser matrix (particularly significant in and near river deltas)	Local effect	
Loss of interstitial fluids (withdrawal of groundwater or oil)	Local effect	? 55 mm/yr
Earthquake-induced vibration	Local effect	
Departure from geoid		
Shifts in hydrosphere, aesthenosphere, core-mantle interface	Local effect	
Shifts in earth's rotation, axis of spin, and precession of equinox	Eustatic	
External gravitational changes	Eustatic	
Evaporation and precipitation (if due to a long-term pattern)	Local effect	

Changes Through Geologic Time

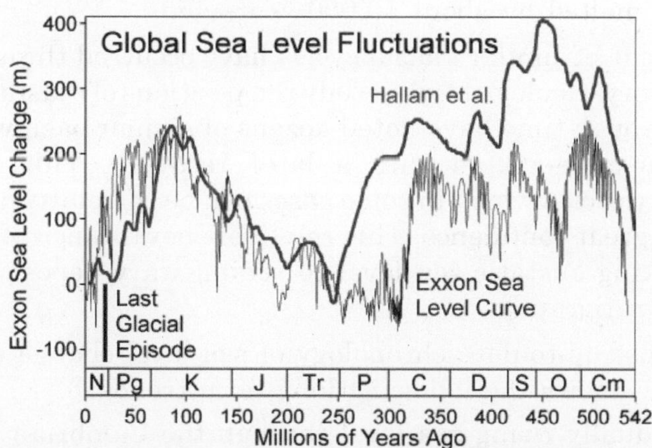

Figure: *Comparison of two sea level reconstructions during the last 500 Ma. The scale of change during the last glacial/interglacial transition is indicated with a black bar. Note that over most of geologic history long-term average sea level has been significantly higher than today.*

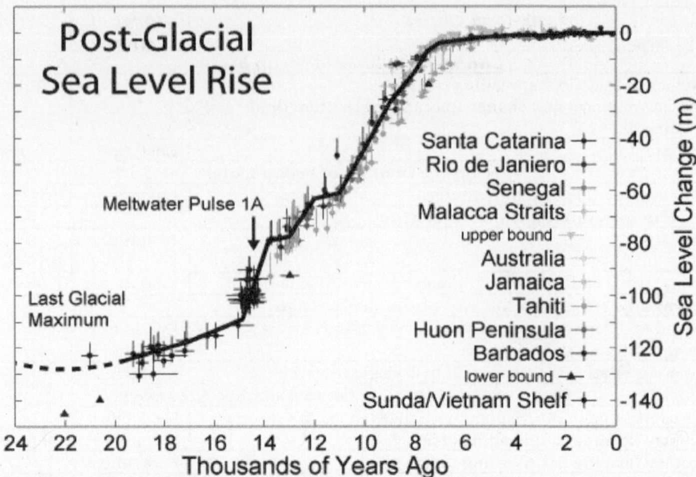

Figure: *Sea level change since the end of the last glacial episode. Changes displayed in metres.*

Sea level has changed over geologic time. As the graph shows, sea level today is very near the lowest level ever attained (the lowest level occurred at the Permian-Triassic boundary about 250 million years ago).

During the most recent ice age (at its maximum about 20,000 years ago) the world's sea level was about 130 m lower than today, due to the large amount of sea water that had evaporated and been deposited as snow and ice, mostly in the Laurentide ice sheet. Most of this had melted by about 10,000 years ago.

Hundreds of similar glacial cycles have occurred throughout the Earth's history. Geologists who study the positions of coastal sediment deposits through time have noted dozens of similar basinward shifts of shorelines associated with a later recovery. This results in sedimentary cycles which in some cases can be correlated around the world with great confidence. This relatively new branch of geological science linking eustatic sea level to sedimentary deposits is called sequence stratigraphy.

The most up-to-date chronology of sea level change during the Phanerozoic shows the following long term trends:

- Gradually rising sea level through the Cambrian
- Relatively stable sea level in the Ordovician, with a large drop associated with the end-Ordovician glaciation
- Relative stability at the lower level during the Silurian

- A gradual fall through the Devonian, continuing through the Mississippian to long-term low at the Mississippian/ Pennsylvanian boundary
- A gradual rise until the start of the Permian, followed by a gentle decrease lasting until the Mesozoic.

Recent Changes

For at least the last 100 years, sea level has been rising at an average rate of about 1.8 mm per year. Most of this rise can be attributed to the increase in temperature of the sea and the resulting slight thermal expansion of the upper 500m of sea water. Additional contributions, as much as one-fourth of the total, come from water sources on land, such as melting snow and glaciers and extraction of groundwater for irrigation and other agricultural and human needs.

Aviation

Using pressure to measure altitude results in two other types of altitude. Distance above *true* or *MSL* (mean sea level) is the next best measurement to absolute. MSL altitude is the distance above where sea level would be if there were no land. If one knows the elevation of terrain, the distance above the ground is calculated by a simple subtraction.

An MSL altitude—called pressure altitude by pilots—is useful for predicting physiological responses in unpressurized aircraft. It also correlates with engine, propeller, and wing performance, which all decrease in thinner air.

Pilots can estimate height above terrain with an altimeter set to a defined barometric pressure. Generally, the pressure used to set the altimeter is the barometric pressure that would exist at MSL in the region being flown over. This pressure is referred to as either QNH or "altimeter" and is transmitted to the pilot by radio from air traffic control (ATC) or an Automatic Terminal Information Service (ATIS). Since the terrain elevation is also referenced to MSL, the pilot can estimate height above ground by subtracting the terrain altitude from the altimeter reading. Aviation charts are divided into boxes and the maximum terrain altitude from MSL in each box is clearly indicated. Once above the transition altitude, the altimeter is set to the international standard atmosphere (ISA) pressure at MSL which is 1013.25 hPa or 29.92 inHg.

Flight Level

MSL is useful for aircraft to avoid terrain, but at high enough altitudes, there is no terrain to avoid. Above that level, pilots are primarily interested in avoiding each other, so they adjust their altimeter to standard temperature and pressure conditions (average sea level pressure and temperature) and disregard actual barometric pressure—until descending below transition level. To distinguish from MSL, such altitudes are called flight levels. Standard terminology is to express flight level as hundreds of feet, so FL 240 is 24,000 feet (7,300 m). Pilots use the international standard pressure setting of 1013.25 hPa (29.92 inHg) when referring to flight levels. The altitude at which aircraft are mandated to set their altimeter to flight levels is called "transition altitude". It varies from country to country. For example in the U.S. it is 18,000 feet, in many European countries it is 3,000 or 5,000 feet.

Current Sea Level Rise

Sea levels around the world are rising. Current sea-level rise potentially affects human populations (e.g., those living in coastal regions and on islands) and the natural environment (e.g., marine ecosystems). Between 1870 and 2004, global average sea levels rose 195 mm (7.7 in). From 1950 to 2009, measurements show an average annual rise in sea level of 1.7 ± 0.3 mm per year, with satellite data showing a rise of 3.3 ± 0.4 mm per year from 1993 to 2009, a faster rate of increase than previously estimated. It is unclear whether the increased rate reflects an increase in the underlying long-term trend.

Two main factors contributed to observed sea level rise. The first is thermal expansion: as ocean water warms, it expands. The second is from the contribution of land-based ice due to increased melting. The major store of water on land is found in glaciers and ice sheets.

Sea level rise is one of several lines of evidence that support the view that the climate has recently warmed. It is very likely that human-induced (anthropogenic) warming contributed to the sea level rise observed in the latter half of the 20th century.

Sea level rise is expected to continue for centuries. In 2007, the Intergovernmental Panel on Climate Change (IPCC) projected that during the 21st century, sea level will rise another 18 to 59 cm (7.1 to 23 in), but these numbers do not include "uncertainties in climate-carbon cycle feedbacks nor do they include the full effects of changes in ice sheet flow". More recent projections assessed by the US National Research Council (2010) suggest possible sea level rise over the 21st century of between 56 and 200 cm (22 and 79 in).

On the timescale of centuries to millennia, the melting of ice sheets 'could result in even higher sea level rise. Partial deglaciation of the Greenland ice sheet, and possibly the West Antarctic ice sheet, could contribute 4 to 6 μ (13 to 20 ft) or more to sea level rise.

Work by a team led by Veerabhadran Ramanathan of the Scripps Institution of Oceanography suggests quick way to stave off impending sea level rise is to cut emissions of short-lived climate warmers such as methane and soot.

Longer-term Changes

Various factors affect the volume or mass of the ocean, leading to long-term changes in eustatic sea level. The two primary influences are temperature (because the density of water depends on temperature), and the mass of water locked up on land and sea as fresh water in rivers, lakes, glaciers, polar ice caps, and sea ice. Over much longer geological timescales, changes in the shape of oceanic basins and in land–sea distribution affect sea level.

Observational and modelling studies of mass loss from glaciers and ice caps indicate a contribution to sea-level rise of 0.2–0.4 mm/yr, averaged over the 20th century.

Glaciers and Ice Caps

Each year about 8 mm of ocean water falls on the Antarctica and Greenland ice sheets as snowfall. If no ice returned to the oceans, sea level would drop 8 mm every year. To a first approximation, the same amount of water appeared to return to the ocean in icebergs and from ice melting at the edges. Scientists previously had estimated which is greater, ice going in or coming out, called the *mass balance*, important because a non-zero balance causes changes in global sea level. High-precision gravimetry from satellites in low-noise flight determined that Greenland was losing more than 200 billion tons of ice per year, in accord with loss estimates from ground measurement. The rate of ice loss was accelerating, having grown from 137 gigatons in 2002–2003. The total global ice mass lost from Greenland, Antarctica and Earth's glaciers and ice caps during 2003-2010 was about 4.3 trillion tons (1,000 cubic miles), adding about 12mm (0.5 inches) to global sea level, enough ice to cover the United States 50cm (1.5 feet) deep. Ice shelves float on the surface of the sea and, if they melt, to a first order they do not change sea level. Likewise, shrinkage/expansion of the northern polar ice cap which is composed of floating pack ice do not significantly affect sea level. Because ice shelf water is fresh, however, melting would cause a very small increase in sea levels, so small that it is generally neglected.

- The melting of small glaciers and polar ice caps on the margins of Greenland and the Antarctic Peninsula melt, would increase sea level around 0.5 m. Melting of the Greenland ice sheet or the Antarctic ice sheet would produce 7.2 μ and 61.1 μ of sea-level rise, respectively. The collapse of the grounded interior reservoir of the West Antarctic Ice Sheet would raise sea level by 5–6 m.
- The interior of the Greenland and Antarctic ice sheets, as of 2009, was sufficiently high (and therefore cold) that direct melt would require several millennia. They could do so through acceleration in flow and enhanced iceberg calving. Also, melt of the fringes of the ice caps could be significant, as could be sub-ice-shelf melting in Antarctica.
- Climate changes during the 20th century were estimated from modelling studies to have led to contributions of between "0.2 and 0.0 mm/yr from Antarctica (the results of increasing precipitation) and 0.0 to 0.1 mm/yr from Greenland (from changes in both precipitation and runoff).
- Estimates suggest that Greenland and Antarctica have contributed 0.0 to 0.5 mm/yr over the 20th century as a result of long-term adjustment to the end of the last ice age.

The current rise in sea level observed from tide gauges, of about 1.8 mm/yr, is within the estimate range from the combination of factors above but active research continues in this field.

In 1992, satellites began recording the change in sea level; they display an acceleration in the rate of sea level change, but they have not been operating for long enough to work out whether this signals a permanent rate change, or an artifact of short-term variation.

Short-term Variability and Long-term Trends

On the timescale of years and decades, sea level records contain a considerable amount of variability. For example, approximately a

10 mm rise and fall of global mean sea level accompanied the 1997–1998 El Niño-Southern Oscillation (ENSO) event, and a temporary 5 mm fall accompanied the 2010–2011 event. Interannual or longer variability is a major reason why no long-term acceleration of sea level has been identified using 20th century data alone.

However, a range of evidence clearly shows that the rate of sea level rise increased between the mid-19th and mid-20th centuries. Sea level acceleration up to the present has been about 0.01 mm/yr^2 and appears to have started at the end of the 18th century.

Sea level rose by 6 cm during the 19th century and 19 cm in the 20th century. Evidence for this includes geological observations, the longest instrumental records and the observed rate of 20th century sea level rise.

For example, geological observations indicate that during the last 2,000 years, sea level change was small, with an average rate of only 0.0–0.2 mm per year. This compares to an average rate of 1.7 ± 0.5 mm per year for the 20th century.

Past Changes in Sea Level

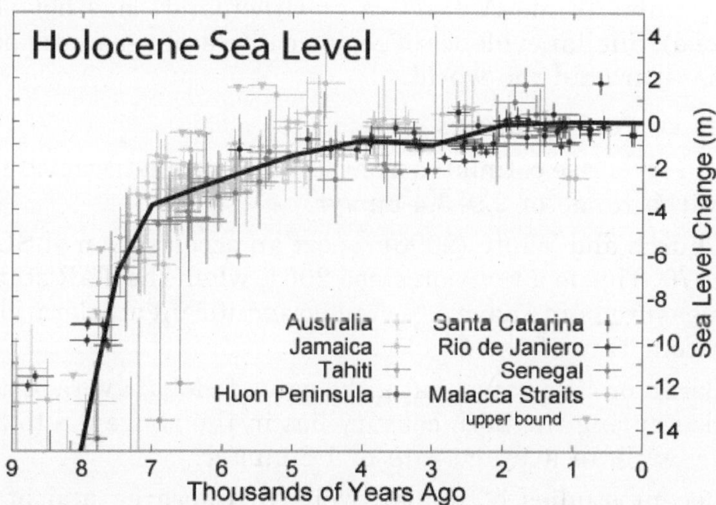

***Figure:** Changes in sea level during the last 9,000 years*

The Sedimentary Record

Sedimentary deposits follow cyclic patterns. Prevailing theories hold that this cyclicity primarily represents the response of depositional processes to the rise and fall of sea level. The rock record indicates that in earlier eras, sea level was both much lower than today and much higher than today. Such anomalies often appear worldwide. For instance, during the depths of the last ice age 18,000 years ago when hundreds of thousands of cubic miles of ice were stacked up on the continents as glaciers, sea level was 120 metres (390 ft) lower, locations that today support coral reefs were left high and dry, and coastlines were miles farther outward.

During this time of very low sea level there was a dry land connection between Asia and Alaska over which humans are believed to have migrated to North America. For the past 6,000 years, the world's sea level gradually approached the current level. During the previous interglacial about 120,000 years ago, sea level was for a short time about 6 metres (20 ft) higher than today, as evidenced by wave-cut notches along cliffs in the Bahamas.

There are also Pleistocene coral reefs left stranded about 3 metres above today's sea level along the southwestern coastline of West Caicos Island in the West Indies. These once-submerged reefs and nearby paleo-beach deposits indicate that sea level spent enough time at that higher level to allow reefs to grow (exactly where this extra sea water came from—Antarctica or Greenland—has not yet been determined). Similar evidence of geologically recent sea level positions is abundant around the world.

Estimates of Past Changes

- Sea level rise estimates from satellite altimetry since 1993 are in the range of 2.9–3.4 mm/yr.
- Church and White (2006) report an acceleration of SLR since 1870. This is a revision since 2001, when the TAR stated that measurements have detected no significant acceleration in the recent rate of sea level rise.
- Based on tide gauge data, the rate of global average sea level rise during the 20th century lies in the range 0.8 to 3.3 mm/yr, with an average rate of 1.8 mm/yr.
- Recent studies of Roman wells in Caesarea and of Roman *piscinae* in Italy indicate that sea level stayed fairly constant from a few hundred years AD to a few hundred years ago.
- Based on geological data, global average sea level may have risen at an average rate of about 0.5 mm/yr over the last 6,000 years and at an average rate of 0.1–0.2 mm/yr over the last 3,000 years.

- Since the Last Glacial Maximum about 20,000 years ago, sea level has risen by more than 120 μ (averaging 6 mm/yr) as a result of melting of major ice sheets. A rapid rise took place between 15,000 and 6,000 years ago at an average rate of 10 mm/yr which accounted for 90m of the rise; thus in the period since 20,000 years BP (excluding the rapid rise from 15–6 kyr BP) the average rate was 3 mm/yr.
- A significant event was Meltwater pulse 1A (mwp-1A), when sea level rose approximately 20 μ over a 500-year period about 14,200 years ago. This is a rate of about 40 mm/yr. The primary source may have been meltwater from the Antarctic ice sheet, perhaps causing the south-to-north cold pulse marked by the Southern Hemisphere Huelmo/Mascardi Cold Reversal, which preceded the Northern Hemisphere Younger Dryas. Other recent studies suggest a Northern Hemisphere source for the meltwater in the Laurentide ice sheet.
- Relative sea level rise at specific locations is often 1–2 mm/yr greater or less than the global average. Along the US mid-Atlantic and Gulf Coasts, for example, sea level is rising approximately 3 mm/yr

US Tide Gauge Measurements

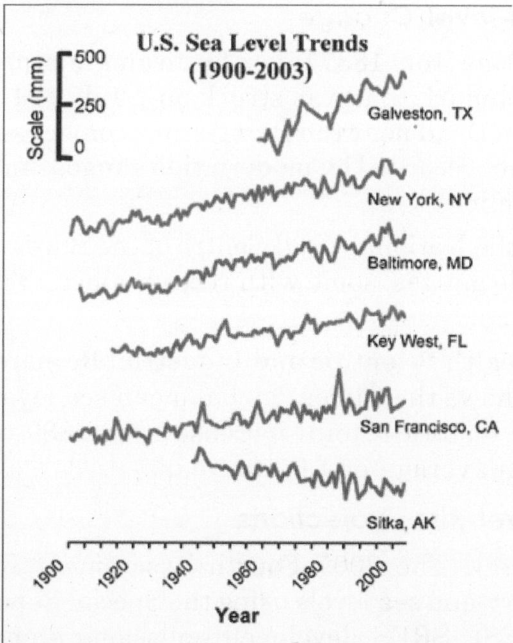

Figure: *US sea-level trends 1900–2003*

Tide gauges in the United States reveal considerable variation because some land areas are rising and some are sinking. For example, over the past 100 years, the rate of sea level rise varied from about an increase of 0.36 inches (9.1 mm) per year along the Louisiana Coast (due to land sinking), to a drop of a few inches per decade in parts of Alaska (due to post-glacial rebound). The rate of sea level rise increased during the 1993–2003 period compared with the longer-term average (1961–2003), although it is unclear whether the faster rate reflected a short-term variation or an increase in the long-term trend.

One study showed no acceleration in sea level rise in US tide gauge records during the 20th century. However, another study found that the rate of rise for the US Atlantic coast during the 20th century was far higher than during the previous two thousand years.

Amsterdam Sea Level Measurements

The longest running sea-level measurements are recorded at Amsterdam, in the Netherlands—part of which (about 25%) lies beneath sea level, beginning in 1700. Since 1850, the rise averaged 1.5 mm/year.

Australian sea-level Change

Records dating from 1837 taken by an amateur meteorologist and a sea level benchmark that was struck on 1 July 1841 on a small cliff on the Isle of the Dead near the Port Arthur convict settlement, when merged with data recorded by modern tide gauges, indicated sea level rise of about 1mm a year.

As of 2003 the National Tidal Centre of the Bureau of Meteorology managed 32 tide gauges, some with records since 1880, for the entire coastline.

Commonwealth Scientific and Industrial Research Organisation (CSIRO) data shows the current global mean sea level trend to be 3.2 mm/yr and the historical total increase from 1880 to 2009 is about 210mm with an average of 1.6 mm/year

Future Sea-level Rise Projections

21st century: The 2007 Fourth Assessment Report (IPCC 4) projected century-end sea levels using the Special Report on Emissions Scenarios (SRES). SRES developed emissions scenarios to project climate-change impacts. The projections based on these scenarios are not predictions, but reflect plausible estimates of future social and economic development (e.g., economic growth, population level). The six SRES "marker" scenarios projected sea level to rise by 18 to 59 centimetres (7.1 to 23 in). Their projections were for the time period 2090–99, with the increase in level relative to average sea level over the 1980–99 period. This estimate did not include all of the possible contributions of ice sheets.

More recent research from 2008 observed rapid declines in ice-mass balance from both Greenland and Antarctica, and concluded that sea-level rise by 2100 is likely to be at least twice as large as that presented by IPCC AR4, with an upper limit of about two metres.

A literature assessment published in 2010 by the US National Research Council described the above IPCC projections as "conservative," and summarized the results of more recent studies. These projections ranged from 56–200 centimetres (22–79 in), based on the same period as IPCC 4.

In 2011, Rignot and others projected a rise of 32 centimetres (13 in) by 2050. Their projection included increased contributions from the Antarctic and Greenland ice sheets. Use of two completely different approaches reinforced the Rignot projection.

After 2100

There is a widespread consensus that substantial long-term sea-level rise will continue for centuries to come. IPCC 4 estimated that at least a partial deglaciation of the Greenland ice sheet, and possibly the West Antarctic ice sheet, would occur given a global average temperature increase of 1–4 °C (relative to temperatures over the years 1990–2000). This estimate was given about a 50% chance of being correct. The estimated timescale was centuries to millennia, and would contribute 4 to 6 metres (13 to 20 ft) or more to sea levels over this period.

There is the possibility of a rapid change in glaciers, ice sheets, and hence sea level. Predictions of such a change are highly uncertain due to a lack of scientific understanding. Modelling of the processes associated with a rapid ice-sheet and glacier change could potentially increase future projections of sea-level rise.

Projected Impacts

Future sea level rise could lead to potentially catastrophic difficulties for shore-based communities in the next centuries: for example, many major cities such as London, New Orleans, and New York already need storm-surge defences, and would need more if the sea level rose, though they also face issues such as subsidence. Sea level rise could also displace many shore-based populations: for example it is estimated that a sea level rise of just 200 mm could create 740,000 homeless people in Nigeria. Maldives, Tuvalu, and other low-lying countries are among the areas that are at the highest level of risk. The UN's environmental panel has warned that, at current rates, sea level would be high enough to make the Maldives uninhabitable by 2100.

Future sea-level rise, like the recent rise, is not expected to be globally uniform (details below). Some regions show a sea-level rise substantially more than the global average (in many cases of more than twice the average), and others a sea level fall. However, models disagree as to the likely pattern of sea level change.

In September 2008, the Delta Commission (Deltacommissie (2007)) presided by Dutch politician Cees Veerman advised in a report that the Netherlands would need a massive new building programme to strengthen the country's water defences against the anticipated effects of global warming for the next 190 years. This commission was created in September 2007, after the damage caused by Hurricane Katrina prompted reflection and preparations. Those included drawing up worst-case plans for evacuations. The plan included more than €100 billion (US$144 bn), in new spending through the year 2100 to take measures, such as broadening coastal dunes and strengthening sea and river dikes.

The commission said the country must plan for a rise in the North Sea up to 4.25 feet (51 inches, 1.3 m) by 2100, rather than the previously projected 2.5 feet (30 inches, 0.80 m), and plan for a 6.5 – 13 feet (80 - 160 inches, 2 – 4 m) rise by 2200.

IPCC Third Assessment

The results from the IPCC Third Assessment Report (TAR) sea level chapter are given below.

The sum of these components indicates a rate of eustatic sea level rise (corresponding to a change in ocean volume) from 1910 to 1990 ranging from "0.8 to 2.2 mm/yr, with a central value of 0.7 mm/yr. The upper bound is close to the observational upper bound (2.0 mm/yr), but the central value is less than the observational lower bound

(1.0 mm/yr), i.e., the sum of components is biased low compared to the observational estimates.

IPCC change factors 1990–2100	***IS92a prediction***	***SRES projection/***
Thermal expansion	110 to 430 mm	
Glaciers	10 to 230 mm (or 50 to 110 mm)	
Greenland ice	? 20 to 90 mm	
Antarctic ice	? 170 to 20 mm	
Terrestrial storage	? 83 to 30 mm	
Ongoing contributions from ice sheets in response to past climate change	0 to 55 mm	
Thawing of permafrost	0 to 5 mm	
Deposition of sediment	not specified	
Total global-average sea level rise (IPCC result, not sum of above)	***110 to 770 mm***	***90 to 880 mm (central value of 480 mm)***

The sum of components indicates an acceleration of only 0.2 (mm/yr)/century, with a range from “1.1 to +0.7 (mm/yr)/century, consistent with observational finding of no acceleration in sea-level rise during the 20th century. The estimated rate of sea-level rise from anthropogenic climate change from 1910 to 1990 (from modelling studies of thermal expansion, glaciers and ice sheets) ranges from 0.3 to 0.8 mm/yr. It is very likely that 20th-century warming has contributed significantly to the observed sea-level rise, through the thermal expansion of sea water and the widespread loss of land ice.

A common perception is that the rate of sea-level rise should have accelerated during the latter half of the 20th century, but tide gauge data for the 20th century show no significant acceleration. Estimates obtained are based on atmosphere-ocean general circulation models (abbreviated AOGCMs) for the terms directly related to anthropogenic climate change in the 20th century, i.e., thermal expansion, ice sheets, glaciers and ice caps... The total computed rise indicates an acceleration of only 0.2 (mm/yr)/century, with a range from “1.1 to +0.7 (mm/yr)/century, consistent with observational finding of no acceleration in sea-level rise during the 20th century. The sum of terms not related to recent climate change is “1.1 to +0.9 mm/yr (i.e., excluding thermal expansion, glaciers and ice caps, and changes in the ice sheets due to 20th century climate change). This range is less than the observational lower bound of sea-level rise. Hence it is very likely that these terms alone are an insufficient explanation, implying that 20th century climate change has made a contribution to 20th century sea-level rise. Recent figures of human, terrestrial impoundment came too late for the 3rd Report, and would revise levels upward for much of the 20th century.

Uncertainty in TAR Sea-level Projections

The different SRES emissions scenarios used for the TAR sea-level projections were not assigned probabilities, and no scenario is assumed by the IPCC to be more probable than another. For the first part of the 21st century, the variation between the different SRES scenarios is relatively small. The range spanned by the SRES scenarios by 2040 is only 0.02 μ or less. By 2100, this range increases to 0.18 m. Of the six illustrative SRES scenarios, A1FI gives the largest sea-level rise and B1 the smallest.

For the TAR sea-level projections, uncertainty in the climate sensitivity and heat uptake of the oceans, as represented by the spread of models (specifically, atmosphere–ocean general circulation models, or AOGCMs), is more important than the uncertainty from the choice of emissions scenario. This differs from the TAR's projections of global warming (i.e., the future increase in global mean temperature), where the uncertainty in emissions scenario and climate sensitivity are comparable in size.

Minority Uncertainties and Criticisms Regarding IPCC Results

- Tide records with a rate of 180 mm/century going back to the 19th century show no measurable acceleration throughout the late 19th and first half of the 20th century. The IPCC attributes about 60 mm/century to melting and other eustatic processes, leaving a residual of 120 mm of 20th-century rise to be accounted for. Global ocean temperatures by Levitus et al. are in accord with coupled ocean/atmosphere modelling of greenhouse warming, with heat-related change of 30 mm. Melting of polar ice-sheets at the upper limit of the IPCC estimates could close the gap, but severe limits are imposed by the observed perturbations in Earth rotation. (Munk 2002)
- By the time of the IPCC TAR, attribution of sea-level changes had a large unexplained gap between direct and indirect estimates of global sea-level rise. Most direct estimates from tide gauges give 1.5–2.0 mm/yr, whereas indirect estimates based on the two processes responsible for global sea-level rise, namely mass and volume change, are significantly below this range. Estimates of the volume increase due to ocean warming

give a rate of about 0.5 mm/yr and the rate due to mass increase, primarily from the melting of continental ice, is thought to be even smaller. One study confirmed tide-gauge data is correct, and concluded there must be a continental source of 1.4 mm/yr of fresh water. (Miller 2004)

- From (Douglas 2002): "In the last dozen years, published values of 20th century GSL rise have ranged from 1.0 to 2.4 mm/yr. In its Third Assessment Report, the IPCC discusses this lack of consensus at length and is careful not to present a best estimate of 20th century GSL rise. By design, the panel presents a snapshot of published analysis over the previous decade or so and interprets the broad range of estimates as reflecting the uncertainty of our knowledge of GSL rise. We disagree with the IPCC interpretation. In our view, values much below 2 mm/yr are inconsistent with regional observations of sea-level rise and with the continuing physical response of Earth to the most recent episode of deglaciation."
- The strong 1997–1998 El Niño caused regional and global sea-level variations, including a temporary global increase of perhaps 20 mm. The IPCC TAR's examination of satellite trends says *the major 1997/98 El Niño-Southern Oscillation (ENSO) event could bias the above estimates of sea-level rise and also indicate the difficulty of separating long-term trends from climatic variability.*

Glacier Contribution

It is well known that glaciers are subject to surges in their rate of movement with consequent melting when they reach lower altitudes and/or the sea. The contributors to Annals of Glaciology, Volume 36 (2003) discussed this phenomenon extensively and it appears that slow advance and rapid retreat have persisted *throughout the mid to late Holocene* in nearly all of Alaska's glaciers. Historical reports of surge occurrences in Iceland's glaciers go back several centuries. Thus rapid retreat can have several other causes than CO2 increase in the atmosphere.

The results from Dyurgerov show a sharp increase in the contribution of mountain and subpolar glaciers to sea-level rise since 1996 (0.5 mm/yr) to 1998 (2 mm/yr) with an average of about 0.35 mm/yr since 1960.

Of interest also is Arendt et al., who estimate the contribution of Alaskan glaciers of 0.14±0.04 mm/yr between the mid-1950s to the mid-1990s, increasing to 0.27 mm/yr in the middle and late 1990s.

Greenland Contribution

Krabill *et al.* estimate a net contribution from Greenland to be at least 0.13 mm/yr in the 1990s. Joughin *et al.* have measured a doubling of the speed of Jakobshavn Isbræ between 1997 and 2003. This is Greenland's largest outlet glacier; it drains 6.5% of the ice sheet, and is thought to be responsible for increasing the rate of sea-level rise by about 0.06 millimetres per year, or roughly 4% of the 20th-century rate of sea-level increase. In 2004, Rignot *et al.* estimated a contribution of 0.04±0.01 mm/yr to sea-level rise from southeast Greenland.

Rignot and Kanagaratnam produced a comprehensive study and map of the outlet glaciers and basins of Greenland. They found widespread glacial acceleration below 66 N in 1996 which spread to 70 N by 2005; and that the ice sheet loss rate in that decade increased from 90 to 200 cubic km/yr; this corresponds to an extra 0.25–0.55 mm/yr of sea level rise.

In July 2005 it was reported that the Kangerdlugssuaq glacier, on Greenland's east coast, was moving towards the sea three times faster than a decade earlier. Kangerdlugssuaq is around 1,000 μ thick, 7.2 km (4.5 miles) wide, and drains about 4% of the ice from the Greenland ice sheet. Measurements of Kangerdlugssuaq in 1988 and 1996 showed it moving at between 5 and 6 km/yr (3.1–3.7 miles/yr), while in 2005 that speed had increased to 14 km/yr (8.7 miles/yr).

According to the 2004 Arctic Climate Impact Assessment, climate models project that local warming in Greenland will exceed 3 °C during this century. Also, ice-sheet models project that such a warming would initiate the long-term melting of the ice sheet, leading to a complete melting of the Greenland ice sheet over several millennia, resulting in a global sea level rise of about seven metres.

Antarctic Contribution

On the Antarctic continent itself, the large volume of ice present stores around 70% of the world's fresh water. This ice sheet is constantly gaining ice from snowfall and losing ice through outflow to the sea. West Antarctica is currently experiencing a net outflow of glacial ice, which will increase global sea level over time. A review of the scientific studies looking at data from 1992 to 2006 suggested a net loss of around 50 gigatons of ice per year was a reasonable estimate (around 0.14 mm of sea-level rise), although significant acceleration of outflow glaciers in the Amundsen Sea Embayment could have more than doubled this figure for the year 2006.

East Antarctica is a cold region with a ground-base above sea level and occupies most of the continent. This area is dominated by small accumulations of snowfall which becomes ice and thus eventually seaward glacial flows. The mass balance of the East Antarctic Ice Sheet as a whole is thought to be slightly positive (lowering sea level) or near to balance. However, increased ice outflow has been suggested in some regions.

In 2011 ice-penetrating radar led to the creation of the first high-resolution topographic map of one of the last uncharted regions of Earth: the Aurora Subglacial Basin, an immense ice-buried lowland in East Antarctica larger than Texas. The map reveals some of the largest fjords or ice cut channels on Earth. Because the basin lies kilometres below sea level, seawater could penetrate beneath the ice, causing portions of the ice sheet to collapse and float off to sea. The map is expected to improve models of ice sheet dynamics.

Sheperd et al. 2012, found that different satellite methods were in good agreement and combing methods leads to more certainty with East Antarctica, West Antarctica, and the Antarctic Peninsula changing in mass by +14 ± 43, –65 ± 26, and –20 ± 14 gigatonnes per year.

Effects of Snowline and Permafrost

The snowline altitude is the altitude of the lowest elevation interval in which minimum annual snow cover exceeds 50%. This ranges from about 5,500 metres above sea-level at the equator down to sea level at about 65° N&S latitude, depending on regional temperature amelioration effects. Permafrost then appears at sea level and extends deeper below sea-level pole-wards. The depth of permafrost and the height of the ice-fields in both Greenland and Antarctica means that they are largely invulnerable to rapid melting. Greenland Summit is at 3,200 metres, where the average annual temperature is minus 32 °C. So even a projected 4 °C rise in temperature leaves it well below the melting point of ice. Frozen Ground 28, December 2004, has a very significant map of permafrost affected areas in the Arctic. The continuous permafrost zone includes all of Greenland, the North of Labrador, NW Territories, Alaska north of Fairbanks, and most of NE Siberia north of Mongolia and Kamchatka. Continental ice above permafrost is very unlikely to melt quickly. As most of the Greenland and Antarctic ice sheets lie above the snowline and/or base of the permafrost zone, they cannot melt in a timeframe much less than several millennia; therefore they are unlikely to contribute significantly to sea-level rise in the coming century.

Polar Ice

The sea level will rise above its current level if more polar ice melts. However, compared to the heights of the ice ages, today there are very few continental ice sheets remaining to be melted. It is estimated that Antarctica, if fully melted, would contribute more than 60 metres of sea level rise, and Greenland would contribute more than 7 metres.

Small glaciers and ice caps on the margins of Greenland and the Antarctic Peninsula might contribute about 0.5 metres. While the latter figure is much smaller than for Antarctica or Greenland it could occur relatively quickly (within the coming century) whereas melting of Greenland would be slow (perhaps 1,500 years to fully deglaciate at the fastest likely rate) and Antarctica even slower. However, this calculation does not account for the possibility that as meltwater flows under and lubricates the larger ice sheets, they could begin to move much more rapidly towards the sea.

In 2002, Rignot and Thomas found that the West Antarctic and Greenland ice sheets were losing mass, while the East Antarctic ice sheet was probably in balance (although they could not determine the sign of the mass balance for The East Antarctic ice sheet). Kwok and Comiso (*J. Climate*, v15, 487–501, 2002) also discovered that temperature and pressure anomalies around West Antarctica and on the other side of the Antarctic Peninsula correlate with recent Southern Oscillation events.

In 2004 Rignot et al. estimated a contribution of 0.04 ± 0.01 mm/yr to sea level rise from South East Greenland. In the same year, Thomas et al. found evidence of an accelerated contribution to sea level rise from West Antarctica. The data showed that the Amundsen Sea sector of the West Antarctic Ice Sheet was discharging 250 cubic kilometres of ice every year, which was 60% more than precipitation accumulation in the catchment areas. This alone was sufficient to raise sea level at 0.24 mm/yr. Further, thinning rates for the glaciers studied in 2002–03 had increased over the values measured in the early 1990s.

The bedrock underlying the glaciers was found to be hundreds of metres deeper than previously known, indicating exit routes for ice from further inland in the Byrd Subpolar Basin. Thus the West Antarctic ice sheet may not be as stable as has been supposed.

In 2005 it was reported that during 1992–2003, East Antarctica thickened at an average rate of about 18 mm/yr while West Antarctica showed an overall thinning of 9 mm/yr. associated with increased precipitation. A gain of this magnitude is enough to slow sea-level rise by 0.12 ± 0.02 mm/yr.

Effects of Sea-level Rise

Based on the projected increases stated above, the IPCC TAR WGII report (*Impacts, Adaptation Vulnerability*) notes that current and future climate change would be expected to have a number of impacts, particularly on coastal systems.

Such impacts may include increased coastal erosion, higher storm-surge flooding, inhibition of primary production processes, more extensive coastal inundation, changes in surface water quality and groundwater characteristics, increased loss of property and coastal habitats, increased flood risk and potential loss of life, loss of non-monetary cultural resources and values, impacts on agriculture and aquaculture through decline in soil and water quality, and loss of tourism, recreation, and transportation functions.

There is an implication that many of these impacts will be detrimental—especially for the three-quarters of the world's poor who depend on agriculture systems. The report does, however, note that owing to the great diversity of coastal environments; regional and local differences in projected relative sea level and climate changes; and differences in the resilience and adaptive capacity of ecosystems, sectors, and countries, the impacts will be highly variable in time and space.

Statistical data on the human impact of sea-level rise is scarce. A study in the April, 2007 issue of *Environment and Urbanization* reports that 634 million people live in coastal areas within 30 feet

(9.1 m) of sea level. The study also reported that about two thirds of the world's cities with over five million people are located in these low-lying coastal areas.

The IPCC report of 2007 estimated that accelerated melting of the Himalayan ice caps and the resulting rise in sea levels would likely increase the severity of flooding in the short term during the rainy season and greatly magnify the impact of tidal storm surges during the cyclone season.

A sea-level rise of just 400 mm in the Bay of Bengal would put 11 percent of the Bangladesh's coastal land underwater, creating 7–10 million climate refugees.

Island Nations

IPCC assessments suggest that deltas and small island states are particularly vulnerable to sea-level rise caused by both thermal expansion and ocean volume.

Sea level changes have not yet been conclusively proven to have directly resulted in environmental, humanitarian, or economic losses to small island states, but the IPCC and other bodies have found this a serious risk scenario in coming decades.

Many media reports have focused on the island nations of the Pacific, notably the Polynesian islands of Tuvalu, which based on more severe flooding events in recent years, were thought to be "sinking" due to sea level rise.

A scientific review in 2000 reported that based on University of Hawaii gauge data, Tuvalu had experienced a negligible increase in sea level of 0.07 mm a year over the past two decades, and that ENSO had been a larger factor in Tuvalu's higher tides in recent years. A subsequent study by John Hunter from the University of Tasmania, however, adjusted for ENSO effects and the movement of the gauge (which was thought to be sinking).

Hunter concluded that Tuvalu had been experiencing sea-level rise of about 1.2 mm per year. The recent more frequent flooding in Tuvalu may also be due to an erosional loss of land during and following the actions of 1997 cyclones Gavin, Hina, and Keli.

Numerous options have been proposed that would assist island nations to adapt to rising sea level.

Satellite sea Level Measurement

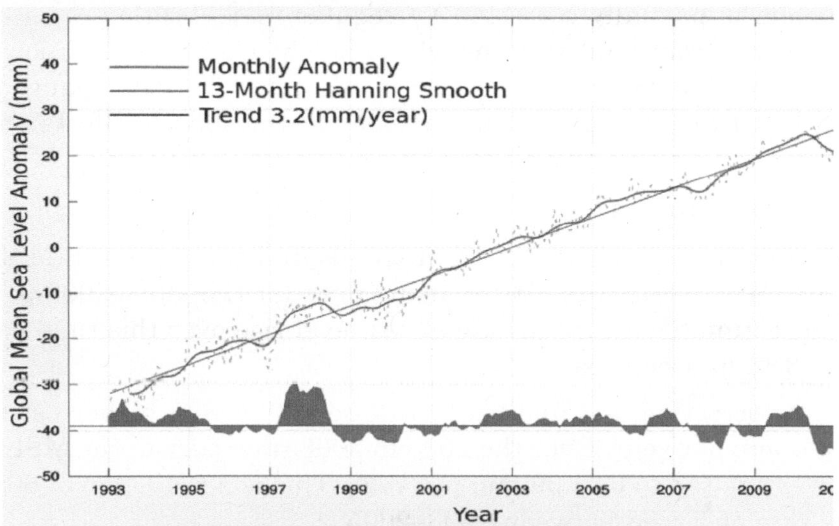

***Figure:** Satellite Measurement of Sea Level.*

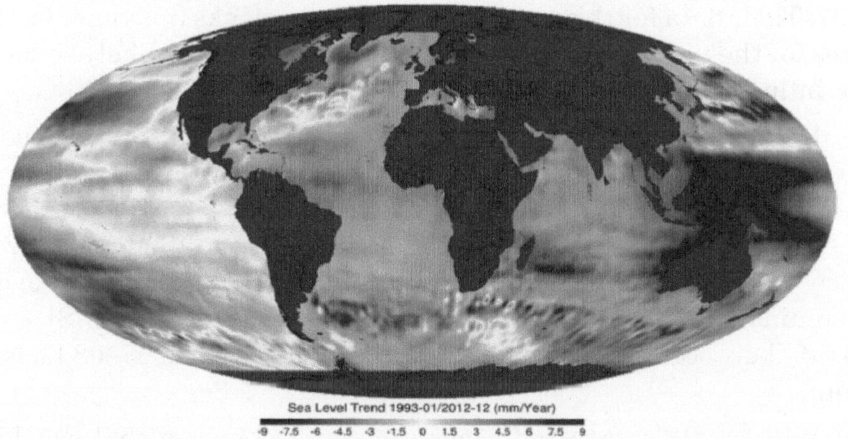

***Figure:** 1993–2012 Sea level trends from satellite altimetry.*

Current rates of sea level rise from satellite altimetry have been estimated in the range of 2.9–3.4 ± 0.4–0.6 mm per year for 1993–2010. This exceeds those from tide gauges. It is unclear whether this represents an increase over the last decades; variability; true differences between satellites and tide gauges; or problems with satellite calibration. Knowing the current altitude of a satellite which can measure sea level to a precision of about 20 millimetres (e.g. the Topex/Poseidon system) is primarily complicated by orbital decay and the difference between the assumed orbit and the earth geoid . This problem is partially corrected by regular re-calibration of satellite altimeters from land stations whose height from MSL is known by surveying. Over water, the height is calibrated from tide gauge data which is needed to correct for tides and atmospheric effects on sea level.

Individual Studies

Ablain *et al.* (2008) looked at trends in mean sea level (MSL). A global MSL curve was plotted using data for the 1993–2008 period. Their estimates for mean rate of sea level rise over this time period was 3.11 mm per year.

A correction was applied to this resulting in a higher estimate of 3.4 mm per year. Over the 2005 to 2008 time period, the MSL rate was estimated to be 1.09 mm per year. This is a reduction of 60% on the rate observed between 1993–2005.

MSL was also plotted using data between the years 1994 and 2007. Their data for this time period show two peaks (maxima) in MSL rates for the years 1997 and 2002. These maxima very likely reflected the influence of the ENSO on MSL.

Using the 1994–2007 MSL data, they estimated MSL rates using moving windows of three and five years. Lower rates were observed during La Niña events in 1999 and 2007. They concluded that the recently observed reduction in the MSL rate was likely to be real, since it coincided with an exceptionally strong La Niña event. Preliminary analyses suggested that an acceleration of the MSL trend would likely occur in relationship with the end of the 2007–08 La Niña event.

White (2011) reported measurements of near-global sea level made using satellite altimeters. Over the time period January 1993 to April 2011, these data show a steady increase in global mean sea level (GMSL) of around 3.2 mm per year, with a range of plus or minus 0.4 mm per year. This is 50% larger than the average rate observed over the 20th century. White (2011) was, however, unsure of whether or not this represented a long-term increase in the rate.

The *Centre National d'Etudes Spatiales/Collecte Localisation Satellites* (CNES/CLS, 2011) reported on the estimated increase in GMSL between 1993 and 2011. Their estimate was an increase of

3.22 mm per year, with an error range in this trend (i.e., the slope over the 1993 to 2011 time period) of approximately 0.6 mm per year.

The CU Sea Level Research Group (CUSLRG, 2011) estimated the rate of GMSL between 1993 and 2011. The rate was estimated at 3.2 mm per year, with a range of plus or minus 0.4 mm per year.

The Laboratory for Satellite Altimetry (LSA, 2011) estimated the trend in GMSL over the time period 1992 to 2011. Their estimate was a trend of 2.9 mm per year, with a range of plus or minus 0.4 mm per year. According to the LSA (2011): "[the] estimates of sea level rise do not include glacial isostatic adjustment effects on the geoid, which are modelled to be +0.2 to +0.5 mm/year when globally averaged."

6

The Effects of Global Change on Soil Conditions

The main potential changes in soil-forming factors (forcing variables) directly resulting from global change would be in organic matter supply from biomass, soil temperature regime and soil hydrology, the latter because of shifts in rainfall zones as well as changes in potential evapotranspiration.

The biggest single change in soils expected as a result of these postulated forcing changes would be a gradual improvement in fertility and physical conditions of soils in humid and subhumid climates.

Other widespread changes would be in degree rather than in kind. Certain tropical soils with low physico-chemical activity, such as in the Amazon region, may undergo a radical change from one major soil-forming process to another.

The changes in temperature but particularly in rainfall to be expected as a result of global warming are subject to major uncertainties for several reasons.

Different global circulation models do not lead to mutually consistent results, and they are not yet adequately verified. Also, the interaction with changes in location and intensity of major ocean currents and resultant possible modifications in sea surface temperatures is still most uncertain, as well as the interaction with possible major changes in cloudiness and land cover and the resulting changes in albedo and actual evapotranspiration.

Indirect effects of climate change on soils through CO_2-induced increases in growth rates or water-use efficiencies, through sea-level rise, through climate-induced decrease or increase in vegetative cover, or a change in human influence on soils because of the changes in options for the farmer, for example, may well each be greater than direct effects on soils of higher temperatures or greater rainfall variability and larger or smaller rainfall totals.

Possible Changes in Forcing Variables

With these caveats, one could stipulate the following changes in forcing variables as likely to materialize sometime during the next century:

- A gradual, continuing rise in atmospheric CO_2 concentration entailing increased photosynthetic rates and water-use efficiencies of vegetation and crops, hence increases in organic matter supplies to soils.
- Minor increases in soil temperatures in the tropics and subtropics; moderate increases and extended periods in which soils are warm enough for microbial activity (warmer than about 5°C) in temperate and cold climates, parallel to the changes in air temperatures and vegetation zones as summarized by Emanuel *et al.*
- Minor increases in evapotranspiration in the tropics to major increases in high latitudes caused both by temperature increase and by extension of the growing period.
- Increases in amount and in variability of rainfall in the tropics; possible decrease in rainfall in a band in the subtropics poleward of the present deserts; minor increases in amount and variability in temperate and cold regions. Peak rainfall intensities could increase in several regions.
- A gradual sea-level rise causing deeper and longer inundation in river and estuary basins and on levee backslopes, and brackish-water inundation leading to encroachment of vegetation that accumulates pyrite in soils near the coast.

This chapter will not touch upon a possibly increased frequency and severity of cyclonic storms in the present cyclone belts or conceivable poleward widening of these belts because of increased sea surface temperatures, which would also give rise to greater frequencies of high-intensity rainfall events.

Effects of Higher CO_2 on Soil Fertility, Physical Conditions and Productivity

Higher atmospheric CO_2 concentration, as discussed in subsequent chapters, increases growth rates and water-use efficiency of crops and natural vegetation in so far as other factors do not become limiting. The higher temperature optima of some plants under increased CO_2 would tend to counteract adverse effects of temperature rise, such as increased nighttime respiration.

The shortened growth cycle of a given species because of higher CO_2 and temperature would be compensated for in natural vegetation by adjustments in species composition or dominance. In agro-ecosystems the choice of longer-duration cultivars or changes in cropping pattern could eliminate unproductive periods that might arise because of the shorter growth cycle of the main crop.

There will be adequate time to adjust to the changes since these are expected to occur over decades, rather than years or days as in all present experimental situations. This heading deals with the effects of *gradually* rising CO_2 concentrations as observed in the recent past and stipulated in simulation models that apply *transient* scenarios.

The increased productivity is generally accompanied by more litter or crop residues, a greater total root mass and root exudation, increased mycorrhizal colonization and activity of other rhizosphere or soil micro-organisms, including symbiotic and root-zone N, fixers. The latter would have a positive effect on N supply to crops or vegetation.

The increased microbial and root activity in the soil would entail higher CO_2 partial pressure in soil air and CO_2 activity in soil water, hence increased rates of plant nutrient release (e.g., K, Mg, micronutrients) from weathering of soil minerals. Similarly, the mycorrhizal activity would lead to better phosphate uptake. These effects would be in synergy with better nutrient uptake by the more intensive root system due to higher atmospheric CO_2 concentration. There is no *a priori* reason why the degree of synchrony between nutrient release and demand by crops or natural vegetation would be subject to major changes under high CO_2 conditions. The greater microbial activity tends to increase the quantity of plant nutrients cycling through soil organisms. The increased production of root material (at similar temperatures) tends to raise soil organic matter content, which also entails the temporary immobilization and cycling of greater quantities of plant nutrients in the soil. Higher C/N ratios in litter, reported by some workers under high CO_2 conditions, would entail slower decomposition and slower remobilization of the plant nutrients from the litter and uptake by the root mat, and would provide more time for incorporation into the soil by earthworms, termites, etc. Higher soil temperatures would counteract increases in 'stable' soil organic matter content but would further stimulate microbial activity.

In all experimental situations, whether chamber-type or free-air enrichment, CO_2 increases are rapid or sudden, often to double ambient concentration, sometimes higher. The consequently rapid increases in soil organic matter dynamics and soil micro-organisms may cause temporary competition for plant nutrients.

These temporary effects have on occasion been reported as negative factors affecting plant response to elevated CO_2 However, increased organic matter dynamics and microbial activity in soils are positive for the soil-plant system when CO_2 concentrations rise gradually over decades, as currently and in the recent past. Future experiments could be set up to compensate for the temporary effects caused by the suddenness of the CO_2 increase, for example by artificially higher soil organic matter contents estimated to be near equilibrium with each stepwise higher CO_2 concentration, in a range between 350 and 600 ppm.

Increased microbial activity due to higher CO_2 concentration and temperature produces greater amounts of polysaccharides and other soil stabilizers. Increases in litter or crop residues, root mass and organic matter content tend to stimulate the activity of soil macrofauna, including earthworms, with consequently improved infiltration rate and bypass flow by the greater number of stable biopores. The greater stability and the faster infiltration increase the resilience of the soil against water erosion and consequent loss of soil fertility. The increased proportion of bypass flow also decreases the nutrient loss by leaching during periods with excess rainfall. This refers to the available nutrients in the soil, including well-incorporated fertilizers or manure, but not to fertilizers broadcast on the soil surface. These are subject to loss by runoff or leaching.

These changes increase the resilience of the soil against physical degradation and nutrient loss by increased intensity, seasonality or variability of rainfall, as well as against some of the unfavourable changes in rate or direction of soil-forming processes discussed in the next sections. If the partial pressure of CO_2 in the soil air would rise, and that of O_2 decrease to levels impairing root function, part of the benefits indicated would not materialize. The improved gas exchange with the atmosphere through increased numbers of stable biopores would tend to keep CO_2 and O_2 in the soil at 'safe' levels, at least in naturally or artificially well-drained soils. Wetland crops such as rice or jute have their own gas exchange mechanisms and would not be affected; neither would natural wetland vegetation.

The positive effect on weathering rate and plant nutrient availability would occur in soils with significant amounts of weatherable minerals, not in very deeply and strongly weathered or otherwise very poor soils.

Effects of Rainfall and Temperature Changes in Different Climates

In the humid tropics and monsoon climates, increased intensities of rainfall events and increased rainfall totals would increase leaching rates in well-drained soils with high infiltration rates, and would cause temporary flooding or water-saturation, hence reduced organic matter decomposition, in many soils in level or depressional sites. This may affect a significant proportion of especially the better soils in Sub-Saharan Africa, for example. They would also give rise to greater amounts and frequency of runoff on soils in sloping terrain, with sedimentation downslope and, worse, downstream. Locally, there would be increased chances of mass movement in the form of landslides or mudflows in certain soft sedimentary materials, discussed below. Soils most resilient against such changes would have adequate cation exchange capacity and anion sorption to minimize nutrient loss during leaching flows, and have a high structural stability and a strongly heterogeneous system of continuous macropores to maximize infiltration and rapid bypass flow through the soil during high-intensity rainfall.

In subtropical and other subhumid or semi-arid areas, the increased productivity and water-use efficiency due to higher CO_2 would tend to increase ground cover, counteracting the effects of higher temperatures. If there would be locally much less rainfall and increasing intra- and inter-annual variability, these could lead to less dry-matter production and hence, in due course, lower soil organic matter contents. Periodic leaching during high-intensity rainfall with less standing vegetation could desalinize some soils in well-drained sites, cause increased runoff in others, and lead to soil salinization in depressional sites or where the groundwater table is high. Soils most resilient against the effects of such increasing aridity and rainfall variability would have a high structural stability and a strongly heterogeneous system of continuous macropores (the same as in the tropics); hence a rapid infiltration rate, as well as a large available water capacity and a deep groundwater table.

Higher temperatures, particularly in arid conditions, entail a higher evaporative demand. Where there is sufficient soil moisture, for example in irrigated areas, this could lead to soil salinization if land or farm water management, or irrigation scheduling or drainage are inadequate. On the other hand, recent experiments by the Salinity Laboratory, Riverside, California, point to increased salt tolerance of crops under high atmospheric CO_2 conditions.

In temperate climates, minor increases in rainfall totals would be expected to be largely taken up by increased evapotranspiration of vegetation or crops at the expected higher temperatures, so that net hydrologic or chemical effects on the soils might be small. The negative effect on soil organic matter contents of a temperature rise might be more than compensated by the greater organic matter supply from vegetation or crops growing more vigorously because of the higher photosynthesis, the greater potential evapotranspiration and the higher water-use efficiency in a high-CO_2 atmosphere. The temperate zone would thus be likely to have the smallest changes in soils, even in poorly buffered ones, directly caused by the effects of global change. A minor and probably slow, but very visible, change could be a reddening of presently brown soils where increased periods with high summer temperatures would coincide with dry conditions, so that the iron oxide haematite would be stable over the presently dominant goethite. This mineralogical change might decrease the intensity and amount of phosphate fixation.

In boreal climates, the gradual disappearance of large extents of permafrost and the reduction of frost periods in extensive belts adjoining former permafrost are expected to improve the internal drainage of soils in vast areas, with probable increases in leaching rates. The appreciable increase in period when the soil temperature is high enough for microbial activity would lead to lower organic matter contents, probably not fully compensated by increased primary production through somewhat higher net photosynthesis and the longer growing period.

Paradoxically, the extent of soils subject to periodic reduction could well increase in level areas, in spite of the greater leaching, because of increased periods when the soils are water-saturated but also sufficiently warm for microbial activity. Soils most resilient against such effects, including the leaching of nutrients and periodic soil reduction, would have similar characteristics as the most resilient ones in other climates: adequate cation exchange capacity and anion sorption to minimize nutrient loss during leaching flows, a high structural stability and a strongly heterogeneous system of continuous macropores to maximize rapid bypass flow during periods with excess meltwater.

Processes in Soils

The most rapid processes of chemical or mineralogical change under changing external conditions would be loss of salts and nutrient cations where leaching increases, and salinization where net upward water movement occurs because of increased evapotranspiration or decreased rainfall or irrigation water supply. The clay mineral composition and the mineralogy of the coarser fractions would generally change little, even over centuries. Exceptions would be the transformation of X-ray amorphous material into the clay mineral halloysite when a volcanic soil previously under perennially moist conditions becomes subject to periodic drying, or the gradual dehydration of goethite to haematite in soils subject to higher temperatures or severe drying, or both. Changes in the surface properties of the clay fraction, while generally slower than salt movement, can take place much faster than changes in bulk composition or crystal structure. Such surface changes have a dominant influence on soil physical and chemical properties.

Changes in the clay mineral surfaces or the bulk composition of the clay fraction of soils are brought about by a small number of transformation processes, listed below. Each of these processes can be accelerated or inhibited by changes in external conditions due to global change.

- hydrolysis by water containing carbon dioxide, which removes silica and basic cations;

- cheluviation, which dissolves and removes especially aluminium and iron by chelating organic acids;
- ferrolysis, a cyclic process of clay transformation and dissolution mediated by alternating iron reduction and oxidation, which decreases the cation exchange capacity by aluminium interlayering in swelling clay minerals;
- dissolution of clay minerals by strong mineral acids, producing acid aluminium salts and amorphous silica;
- reverse weathering, i.e., clay formation and transformation under neutral to strongly alkaline conditions, which may create, e.g., montmorillonite, palygorskite or analcime.

Hydrolysis and cheluviation may be accelerated by increased leaching rates. Ferrolysis may occur where soils are subject to reduction and leaching in alternation with oxidation: in a warmer world, this may happen over larger areas than at present, especially in high latitudes and in monsoon climates. Dissolution by strong acids would occur, e.g., where sulphidic materials in coastal plains are oxidized with an improvement of drainage; however, a rise in sea level would reduce the likelihood of this occurring naturally. Reverse weathering could begin in areas drying out during global warming, and would continue in most presently arid areas.

These processes would influence the surface properties of the clay fraction only over a period of centuries, even with the changes envisaged as a consequence of global warming. By contrast, direct human action can vastly accelerate some of these processes as is evident, for example, from the severe effects of acid rain on sandy soils in parts of Europe or from the extremely rapid ferrolysis in soils seasonally inundated by water level fluctuations in the Volta lake in Ghana.

Not only the speed of soil formation can be accelerated by human action, but also, albeit much more locally, its very nature or direction. In most places, the natural soil-forming processes are not fundamentally changed, but there are certain threshold situations, generally with fragile soils, where even a small change in external conditions may cause a major, and adverse, change from one dominant soil-forming process to another. The yellowish sandy Ferralsols and Ferralic Arenosols of Eastern Amazonia, Kalimantan and the Zaire basin may rapidly change into Podzols or Albic Arenosols (giant Podzols) with even small increases in total rainfall or stronger seasonality, or increased input of acidic ('poor') organic matter. An increase in effective rainfall due to climate change may cause a major increase in the extent of Podzols formed from present-day yellowish sandy Ferralsols where, presently, Podzols occur in patches within the Ferralsols area.

The imperfectly drained loamy Plinthosols on the flat interfluves of Western Amazonia would change into shallow, droughty soils with an irreversibly hardened subsoil if subject to drying out with climate change.

The deep reddish, porous loamy to clayey Ferralsols of the transition zones between forest and savanna in Eastern Africa, stable under the present vegetation, may be leached so far that a denser subsoil with washed-in clay is formed below an unstable topsoil with little organic matter, as already observed where the land was cleared several decades ago; the same may happen over more extensive areas under a sparser vegetation brought about by a somewhat drier climate.

The silty Fluvisols in the broad river valleys of the Sudano-Sahelian zone of West Africa, such as the interior delta of the Niger river, may become saline or sodic upon even minimal change in precipitation and flooding regimes - as exemplified by current human actions with the same soil-hydrological implications.

Some Properties of Clay Surfaces

Soils with a naturally high structural stability, for example Ferralsols and Nitisols, occupy sizeable areas in the tropics. The former are widespread, among others, in northern South America, the latter in East Africa. The clay fraction in such soils generally has oxidic surfaces: mainly iron(III) and aluminium oxides or hydroxides, while the bulk of the clay fraction may have different compositions. The oxidic surfaces could form from parent materials with moderate or high iron contents under long-continued hydrolysis by water (containing carbon dioxide).

At the other extreme are soils with a very low structural stability, or with a severe hazard of failure under load or shock (so-called quick clays). The surfaces of the clay minerals in these soils are generally covered by amorphous, gel-like material with a high silica content. Such material may have originated in earlier periods when the soils were strongly saline and, presumably, subject to processes of reverse weathering. Examples are the quick clays of the Champlain Sea sediments in Ontario and Quebec and of parts of Scandinavia. Such soils are most likely to generate high proportions of runoff and suspended sediment, but also most liable to mudflows once sloping sites are water-saturated to appreciable depth. Some Andosols are thixotropic and have similarly low stability because of their similar composition, derived from volcanic materials (tuff).

Most soils fall somewhere between these extremes. Vertisols, for example, have moderate or low structural stability, and clay surfaces that are mainly silica, but with generally small amounts of amorphous coating. In Planosols, if formed by ferrolysis, the clay fraction in the upper, eluvial horizons has been partly decomposed with a residue of amorphous silica, but the remaining smectite or illite has been interlayered with aluminium hydroxide polymers, which has decreased the swell-shrink potential and the cation exchange capacity of the clay fraction. Concurrently, parts of the free iron oxides have been reduced and leached out. The net effect of these changes generally is a decrease in structural stability.

Resilience Against Physical and Chemical Soil Degradation

As discussed, most soils do not have a high intrinsic resilience against physical soil degradation by, for example, high-intensity rainfall. In natural conditions in humid climates, it is the complete soil cover near ground level combined with the perforating activity of the soil fauna that makes the soil-vegetation system resilient against physical degradation.

In the Rhine river plain in the Netherlands, for example, most of the originally calcareous alluvial soils have been decalcified within a millennium or so. Only in small areas on the highest levees of that age that have continually remained under forest, soils are still calcareous, and even have lime pseudomycelia (filaments) indicative of less humid soil conditions, and abundant vertical macropores produced by earthworms. In these soils, faunal activity is high because of the adequate litter supply: the resulting macropores remain open, protected against rain impact by litter and undergrowth: and excess water from heavy rain passes to the substratum through the macropores without leaching lime from most of the soil mass.

Research to increase resilience of soils against any adverse effects of climate change could be done in conjunction with that on soil resilience against direct adverse human impacts. Until site-specific management procedures have been elaborated, soil and crop (including trees and pasture) management should aim to maintain soil cover and organic matter supply to soil biota, while minimizing mechanical disturbance by heavy traffic, cultivation or excessive grazing intensity. Such kind of management may also help to conserve plant nutrients (in soils not flooded for wetland cultivation) since the stable, heterogeneous system of biopores produced by the soil fauna would favour bypass flow of any excess moisture and thus decrease leaching through the soil mass.

A single management recipe would not be generally applicable in different conditions. Minimizing damage by certain termite species harming crop performance may necessitate a period without residues on the soil, for example; or crop residues may be needed for feed or fuel. Management methods for wetland need to be developed that make optimum use of any increased potential productivity, while minimizing secondary effects such as increased CH_4 or N_2O emission from the reduced soil. Such factors, and others, should be taken into account in designing an optimum management strategy for any specific natural and cultural environment.

Resilience Against Soil Reduction (Anoxic Conditions)

Soil reduction, which would limit land suitability for dryland crops, or strong reduction, which would be liable to produce toxins even for wetland crops, may take place once the soil is water-saturated long enough for microbial action to exhaust the oxygen remaining in the soil when water-saturation started. Another necessary condition is the presence of sufficient readily decomposable organic matter as an energy source for the microbial activity. In most soils, during reduction the redox status is stabilized at an Eh about 100-200 mV near neutrality by the Fe^{2+} - $Fe(OH)_3$ equilibrium, except where the content of readily decomposable organic matter is very high or the content of free iron (III) oxides very low. In such cases, negative Eh values may occur, and toxic hydrogen sulphide or low-molecular organic compounds - including methane - may be formed.

Resilience against soil reduction in practice depends on the drainage conditions, since most soils have sufficient organic matter for reduction to start within about a week after water-saturation. Soils most resilient against reduction in conditions of increased rainfall variability and incidence of high-intensity rainfall have similar properties as those resilient against the negative effects of other perturbations: high infiltration rate, high structural stability and a permanent heterogeneous system of tubular macropores, good external drainage.

Soil Reaction

Most soils would not be subject to rapid pH changes resulting from climate change. Exceptions might be found in potential acid sulphate soils, extensive in some coastal plains and estuaries, if they become subject to increasingly long dry seasons.

Even though most of such soils are clays with moderate or high cation exchange capacity, the amounts of acid liberated in such soils upon oxidation generally exceed this rapid buffering capacity. Therefore, pH values may temporarily reach 2.5 to 3.5 and a small part of the clay fraction may be decomposed as indicated under *Processes in soils,* above. This then buffers the pH generally between 3.5 and 4 in the long run. Depending on the efficiency with which the excess acid formed can be leached out, the period of extreme acidity and aluminium toxicity may last between less than a year and several decades.

In calcareous soils, soil reaction may range between about 8.5 and 7 depending on the partial pressure of CO_2 in the soil; this range is maintained against leaching of basic cations by the different soil processes as long as a few per cent of finely distributed lime remain. Buffering in non-calcareous soils is less strong, but depends on the cation exchange capacity at soil pH. In soils with variable-charge surfaces of the clay fraction, this decreases with acidification.

It should be noted that the simple modelling of accelerated $CaCO_3$ leaching under a doubled atmospheric CO_2 concentration generally does not hold true. In most soils, the ongoing decomposition of organic matter maintains CO_2 concentrations in the soil air far above atmospheric concentration even now, and $CaCO_3$ solubility is determined by the partial pressure of CO_2 in soil air and its activity in soil water, rather than in the atmosphere. Leaching of lime is thus positively related to rate of organic matter decomposition, negatively to gas diffusion rate, and positively to amount of water percolating through the soil.

In conditions where leaching is accelerated by climate change, it would be possible to find relatively rapid soil acidification after a long period with little apparent change, as has been the case - but after a shorter latent period - in some soils in Europe that have been subject to acid rain for several decades. The soil might in fact be steadily depleted of basic cations, but a pH change may start, or may become more rapid, once certain buffering pools are nearly exhausted. Such non-linear and time-delayed effects have been discussed in the context of soil and water pollution by Stigliani (1988); they are also expected to occur in various ways at different times after increased temperatures and changed rainfall patterns will have been operative.

Effects of a Rising Sea Level on Soils in Coastal Areas

The probable effects on soil characteristics of a gradual eustatic rise in sea-level will vary from place to place depending on a number of local and external factors, and interactions between them. In principle, a rising sea level would tend to erode and move back existing coastlines.

However, the extent to which this actually happens will depend on the elevation, the resistance of local coastal materials, the degree to which they are defended by sediments provided by river flow or longshore drift, the strength of longshore currents and storm waves, and on human interventions which might prevent or accelerate erosion.

In major deltas, such as those of the Ganges-Brahmaputra and the major Chinese rivers, sediment supplies delivered to the estuary will generally be sufficient to offset the effects of a rising sea level. Such deltaic aggradation could decrease, however, under three circumstances:

- where human interventions inland, such as large dams or successful soil conservation programmes, drastically reduce sediment supply to the delta: e.g., the construction of the Aswan high dam in 1964 has led to coastal erosion and increased flooding of lagoon margins in the Nile delta;
- where construction of embankments within the delta interrupts sediment supply to adjoining backswamps, exposing them to submergence by a rise in sea level: e.g., embankments along the lower Mississippi river have cut off sediment supplies to adjoining wetlands which formerly offset land subsidence occurring due to compaction of underlying sediments;
- where land subsidence occurs due to abstraction of water, natural gas or oil: e.g., as is presently happening in Bangkok and in the northern part of the Netherlands.

In coastal lowlands which are insufficiently defended by sediment supply or embankments, tidal flooding by saline water will tend to penetrate further inland than at present, extending the area of perennially or seasonally saline soils. Where *Rhizophora* mangrove or *Phragmites* vegetation invades the area, that would over several decades lead to the formation of potential acid sulphate soils.

Impedance of drainage from the land by a higher sea level and by the correspondingly higher levels of adjoining estuarine rivers and their levees, will also extend the area of perennially or seasonally reduced soils and increase normal inundation depths and durations in river and estuary basins and on levee backslopes.

In sites which become perennially wet, soil organic matter contents will tend to increase, resulting eventually in peat formation. On the other hand, where coastal erosion removes an existing barrier of mineral soils or mangrove forest, higher storm surges associated with a rising sea level could allow sea-water to destroy existing coastal eustatic peat swamps, which might eventually be replaced by freshwater or saltwater lagoons.

The probable response of low-lying coastal areas to a rise in sea level can be estimated in more detail on the basis of the geological and historical evidence of changes that occurred during past periods when sea level was rising eustatically or in response to tectonic or isostatic movements: e.g., around the southern North Sea; in the Nile delta; on the coastal plain of the Guyanas; in the Musi delta of Sumatera. Contemporary evidence is available in areas where land levels have subsided as a result of recent abstraction of water, natural gas or oil from sediments underlying coastal lowlands. Further studies of such contemporary and palaeoenvironments are needed together with location-specific studies in order to better understand the change processes, identify appropriate responses and assess their technical, ecological and socio-economic implications.

Conclusions

Some major and widespread soil changes expected as a result of any global change are positive, especially the gradual increases in soil fertility and physical qualities consequent on increased atmospheric CO_2 The increased productivity and water-use efficiency of crops and vegetation, and the generally similar or somewhat higher rainfall indicated by several global circulation models, not fully counteracted by higher evapotranspiration, would be expected to lead to widespread increases in ground cover, and consequently better protection against runoff and erosion.

Major but less widespread soil changes, including greater biological activity and increased extent of periodic reduction in soils, would be expected where permafrost would disappear. In unprotected low-lying coastal areas, gradual encroachment of *Rhizophora* mangroves or *Phragmites* following more extensive brackish-water inundation may give rise to the formation of potential acid sulphate soil layers after several decades. Deeper and longer-duration flooding of basins and levee backslopes in adjacent river and estuary plains could lead to more extensive reducing conditions and increased organic matter contents, and locally to peat formation.

Other changes due to climate change (temperature and precipitation) are expected to be relatively well buffered by the mineral composition, the organic matter content or the structural stability of many soils. However, decreases in cover by vegetation or annual or perennial crops, caused by any locally major declines in rainfall not compensated by CO_2 effects, could lead to soil structure degradation and decreased porosity, as well as increased runoff and erosion on sloping sites and by the concomitant more extensive and rapid sedimentation. Changes in options available to land users because of climate change may have similar effects.

In certain fragile soils, the nature of the dominant soil-forming process may change for the worse with increased, decreased or more strongly seasonal rainfall. In most cases, changes in soils by direct human action, on-site or off-site (whether intentional or unintended), are far greater than the direct climate-induced effects. Soil management measures designed to optimize the soil's sustained productive capacity would therefore be generally adequate to counteract any degradation of agricultural land by climate change. Soils of nature areas, or other land with a low intensity of management such as semi-natural forests used for extraction of wood and other products, are less readily protected against the effects of climate change but such soils, too, are threatened less by climate change than by human actions - off-site, such as pollution by acid deposition, or on-site, such as excessive nutrient extraction under very low-input agriculture.

To armour the world's soils against any negative effect of climate change, or against other extremes in external circumstances such as nutrient depletion or excess (pollution), or drought or high-intensity rains, the best that land users could do, would be:

- to manage their soils to give them maximum physical resilience through a stable, heterogeneous pore system by maintaining a closed ground cover as much as possible;
- to use an integrated plant nutrient management system to balance the input and off take of nutrients over a cropping cycle or over the years, while maintaining soil nutrient levels low enough to minimize losses and high enough to buffer occasional high demands.

An analogous philosophy, at lower levels of external inputs, could be formulated for extensive grazing land and production forest, whether planted or managed natural forest. Human action and management has been emphasized in these conclusions because most of the world's land is used and, to different degrees, managed rather than under natural conditions.

The CO_2 Fertilization Effect: Higher Carbohydrate Production and Retention as Biomass and Seed Yield

The rise in atmospheric carbon dioxide (CO_2) concentration from about 280 μ mol/mol before the industrial revolution to about 360 μ mol/mol currently is well documented. The consensus of many studies of the effects of elevated CO_2 on plants is that the CO_2 fertilization effect is real. However, the CO_2 fertilization effect may not be manifested under conditions where some other growth factor is severely limiting, such as low temperature (Long, 1991). Also, plants grown in some conditions, where limitations of rooting volume, light, or other factors restrict growth, have not shown a sustained response to elevated CO_2.

The main objectives of this chapter are to assess the direct effects of rising atmospheric CO_2 and indirect effects of potential climate changes on crop growth and yield. The approach will be to (1) provide a general overview of CO_2 effects on plant growth processes; (2) analyse some specific experimental data on crop plant responses to elevated CO_2 and climate change factors; (3) summarize recent reviews of plant responses to elevated CO_2; and (4) discuss some crop modelling assessments of rising CO_2 and climate change factors on agricultural productivity based on predictions of global climate change models.

Also, some adaptations for improving crop productivity in a higher CO_2 world will be suggested.

Overview of CO_2 Effects on Plant Growth Processes

Most of the following discussion of CO_2 effects on plants applies to species with the C_3 photosynthetic pathway and not necessarily to species with the C_4 pathway. Other aerial, non-biotic environmental factors that affect plant growth and development are light and temperature. Plant photosynthetic rates generally increase linearly with light across relatively low ranges of light intensity, and then the rates decelerate until they reach an asymptotic maximum.

Because of crowding and shading of many leaves, most crop canopies do not reach light saturation at full sunlight; that is, they would be able to respond to light levels well beyond full solar irradiance. Likewise, crop photosynthetic rates respond to increasing levels of CO_2 but then level off at higher concentrations (around 700 μ mol/mol or greater, depending upon species and other factors).

However, leaf photosynthesis usually increases with temperature up to some maximum value, and then declines. Furthermore, temperature affects not only photosynthesis, but also respiration, growth, development phases and reproductive processes. Elevated CO_2 may have some effects on crop phenology, although stages of development are governed primarily by temperature, time and photoperiod. If dates of planting were to be changed because of the greenhouse effect, then phenological timing of plants could be affected. For example, higher temperatures could decrease yields by decreasing the duration of the grain-filling period or changes in photoperiod could shorten or lengthen the vegetative stage.

The CO_2 fertilization effect begins with enhanced photosynthetic CO_2 fixation. Non-structural carbohydrates tend to accumulate in leaves and other plant organs as starch, soluble carbohydrates or polyfructosans, depending on species. In some cases, there may be feedback inhibition of photosynthesis associated with accumulation of non-structural carbohydrates. Increased carbohydrate accumulation, especially in leaves, may be evidence that crop plants grown under CO_2 enrichment may not be fully adapted to take complete advantage of elevated CO_2. This may be because the CO_2-enriched plants do not have an adequate sink (inadequate growth capacity), or lack capacity to load phloem and translocate soluble carbohydrates. Improvement of photoassimilate utilization should be one goal of designing cultivars for the future.

In the process of growth, photoassimilates are allocated to the vegetative shoots, root system or reproductive organs. In some cases, more photoassimilate of CO_2-enriched plants is partitioned to the root system than to the shoots. Above ground, more photoassimilate usually goes into stems and supporting structures than into leaves. This phenomenon may not be an inherent response to elevated CO_2, but may be a by-product of the larger size of plants often found in CO_2-enriched atmospheres, especially by species that produce branch stems along the aerial mainstems.

Reproductive biomass growth as well as vegetative biomass growth are usually increased by elevated CO_2. However, the harvest index, or the ratio of seed yield to above-ground biomass yield, is typically lower under elevated CO_2 conditions, which may also be evidence of the lack of capacity to utilize completely the more abundant photoassimilate. In many cases, both the amount and the carboxylation activity of ribulose 1,5-bisphosphate carboxylase-oxygenase enzyme (rubisco) is decreased in leaves of plants grown under elevated CO_2 This acclimation phenomenon may produce 'downregulation of photosynthesis'; however, this is not universally the case. For example, there is little evidence of this downregulation response in soybean (*Glycine max* L. Merr.), a C_3 legume. In fact, photosynthetic capacity per unit leaf area of soybean is increased under CO_2 enrichment. Leaves often develop an additional layer of mesophyll cells. Also, more structural carbohydrate may be produced in leaves, as well as stems, of CO_2-enriched plants.

We can obtain clues about reasons that some plants downregulate and others do not in response to elevated CO_2 by focusing on the capabilities of soybean. This plant has (a) symbiotic N2 fixation; (b) the capacity to form additional layers of palisade cells in the leaf tissue; (c) the capacity to shunt much of the photoassimilate into relatively inert starch rather than soluble sugars during photosynthesis; (d) a relatively strong leaf and stem sink during vegetative development; and (e) a strong seed-fill sink during reproductive development. Plants which lack these capacities, either inherently or because of growth in limiting environments, are more likely to demonstrate some degree of downregulation of photosynthesis.

The carbon:nitrogen ratio of leaves of plants is usually increased under CO_2 enrichment. Plants may acclimate to elevated CO_2 by requiring less rubisco and photo-synthetic apparatus, which would lead to lower nitrogen contents. The overall change in C:N ratios is governed both by increases in structural and non-structural carbohydrates, and by decreases in protein content. However, seed nitrogen content is little affected.

Specific respiration rates may be reduced by both short-term exposure to elevated CO_2 and long-term growth at elevated CO_2. However, the long-term effect may be similar when respiration rates are reported on a per unit nitrogen basis.

In climate change scenarios, temperatures are predicted to increase following the rise of CO_2 and other greenhouse-effect gases. Carbon dioxide x temperature interactions have been observed for vegetative growth (i.e., the CO_2 fertilization effect is greater at warmer temperatures than at cooler temperatures). Temperature increases in a higher CO_2 world could increase overall biomass productivity for vegetative crops (pastures and forages) both by extending the length of the growing season in temperate regions, and by the interaction of CO_2 x temperature in stimulation of vegetative growth. However, CO_2 x temperature interactions appear to be very small or negligible for reproductive processes (seed set and seed yield) although there may be more initial flowers formed by greater amounts of branching or tillering that is stimulated by CO_2 enrichment. Precipitation changes may occur along with other climatic change effects. In general, predictions from crop models show that increased CO_2 should increase productivity of C_3 plants, but the associated predictions of temperature rise will be detrimental. Not surprisingly, changes in precipitation patterns (decreases of rainfall during growth period) could be more detrimental for crop production than changes in temperature.

Under elevated CO_2 stomatal conductance in most species will decrease which may result in less transpiration per unit leaf area. However, leaf area index of some crops may also increase. The typical 40% reduction in stomatal conductance induced by a doubling of CO_2 has generally resulted in only a 10% (or less) reduction in crop canopy water use in chamber or field experimental conditions. Actual changes in crop evapotranspiration will be governed by the crop energy balance, as mitigated by stomatal conductance, leaf area index, crop structure and any changing meteorological factors.

Water-use efficiency (WUE) (ratio of CO_2 uptake to evapotranspiration) will increase under higher CO_2 conditions. This increase is caused more by increased photosynthesis than it is by a reduction of water loss through partially closed stomata. Thus, more biomass can be produced per unit of water used, although a crop would still require almost as much water from sowing to final harvest. If temperatures rise, however, the increased WUE caused by the CO_2 fertilization effect could be diminished or negated, unless planting dates can be changed to more favourable seasons.

Several assessments of impacts of climate change on crop productivity have been published. Progress has been made on integrating the impacts on individual countries and on economic and social interactions. For the most part, these assessments project more favourable climates for agriculture in northern latitudes and less favourable climates in the tropical and subtropical zones. However, the crop modelling predictions are dependent on the scenarios of outputs of General Circulation Models (GCMs) applied to the greenhouse effect.

Thus, the dependency chain of assessments follows: Climate Change Scenarios ® Crop Model Prediction and Agricultural Production Systems (with and without available mitigation and adaptation response strategies) ® National Scenarios of Economics and Well-being of Farmers, Agricultural Commerce, and Consumers ® Country-by-Country and Global Interaction Scenarios of World Trade (food and all other commodities). Population Dynamics and Economic Well-being, and Impacts on Social Systems. As the world continues to consume fossil fuels, CO_2 concentrations will continue to rise. Other greenhouse-effect gases, such as methane, nitrous oxides, chlorofluorocarbons and chlorofluorocarbon substitutes, and perhaps tropospheric ozone, will likely rise also. The CO_2 fertilization effect on plants will increase and climate changes may occur because of the combined increase of all greenhouse-effect gases. Global agriculture could adapt to gradual regional climate changes, but sudden changes would be more serious. Adaptation and/or mitigation actions could include the following:

1. Selection of plants that can better utilize carbohydrates which are produced when plants are grown at elevated CO_2.
2. Selection of plants that produce less structural matter and more reproductive capacity under CO_2 enrichment. (This applies for seed crop plants, not necessarily vegetative biomass plants.)

3. Search for germplasms that are adapted to higher day and night temperatures, and incorporate those traits into desirable crop production cultivars to improve flowering and seed set.
4. Change planting dates and other crop management procedures to optimize yields under new climatic conditions, and select for cultivars that are adapted to these changed agricultural practices.
5. Shift to species that have more stable production under high temperatures or drought.
6. Determine whether more favourable N:C ratios can be attained in forage cultivars adapted to elevated CO_2.
7. Where needed, and where possible, develop irrigation systems for crops.

Specific Responses of Crops to Elevated CO_2

This section will focus on responses of two C_3 crops to elevated CO_2: soybean, a symbiotic nitrogen-fixing legume representative of the pulses; and rice (*Oryza sativa* L.), a globally important cereal food crop. Findings from seven crop cycles of both soybean and rice grown in sunlit chambers at Gainesville, Florida, USA, will be emphasized (Baker and Allen, 1993a,b). The chambers feature both large rooting volumes of real soil and light from the sun. Some studies included responses to sub-ambient as well as superambient CO_2 concentrations.

Soybean

Photosynthetic Rates: Soybean photosynthetic rates of both leaves and the whole canopy have *always* been increased by elevated CO_2 Jones *et al.* (1984) reported midday maximum canopy photosynthetic rates of 60 and 90 μ mol $CO_2/m^2/s$ at 70 days after planting (DAP) for soybean grown at CO_2 concentrations of 330 and 800 μ mol/mol, respectively. The leaf area index (LAI) was 6.9 and 9.0, respectively. Midday maximum canopy photo-synthetic rates in other studies were 40 and 75 μ mol/m^2/s for soybean grown at 320 and 640 μ mol/mol CO_2 and 40 and 80 μ mol/m^2/s for soybean grown at 330 and 800 μ mol/m^2/s.

Valle *et al.* found that midday maximum photosynthetic CO_2 uptake rates of soybean leaves ranged from 30 to 50 μ mol/m^2/s and 15 to 25 μ mol/m^2/s on plants grown at 660 and 330 μ mol/mol CO_2 respectively. Allen *et al.* reported that, at all light levels, leaf photosynthetic rates increased linearly with CO_2 concentration across the range of 330 to 800 μ mol/mol.

Valle *et al.* used a Michaelis-Menten type of rectangular hyperbola to summarize photosynthetic responses of soybean leaves vs. CO_2 concentration. The plants had been grown at 330 and 660 μ mol/mol of CO_2 and then exposed to a wide range of CO_2 for a short period.

$$Y=(Y_{max} \times [C])/([C]+K_m)+Y_i \quad (4.1)$$

where Y is photosynthetic rate in μ mol $CO_2/m^2/s$; $[C_3]$ is CO_2 concentration in μ mol/mol; Y_i is the y-axis intercept at zero $[C_3]$, the apparent respiration rate, in μ mol $CO_2/m^2/s$; Y_{max} is the response limit of $(Y - Y_i)$ at very high [C], the asymptotic photosynthetic rate, in μ mol $CO_2/m^2/s$; K_m is the value of [C] where $(Y-Y_i)=Y_{max}/2$, the apparent Michaelis-Menten constant, in (m mol/mol; and G_c is the calculated [C] intercept at zero Y, the CO_2 compensation point, μ mol/mol (not shown in this equation). The average parameters for responses at 330 and 660 μ mol/mol. There was no obvious downregulation of soybean leaf photosynthesis in response to elevated CO_2; in fact, photosynthetic capacity was increased. Leaf quantum yield increased from 0.05 to 0.09 in the soybean leaves exposed to CO_2 of 330 and 660 μ mol/mol, respectively.

Campbell *et al.* (1988) showed that soybean leaf photosynthetic rates were higher for plants grown at 660 than at 330 μ mol/mol CO_2 when measured at common intercellular CO_2 concentrations. Furthermore, Campbell *et al.* (1988) measured rubisco activity and amount in leaves of soybean grown in CO_2 concentrations of 160, 220, 280, 330, 660 and 990 μ mol/mol. They found that rubisco activity was almost constant at 1.0 μ mol CO_2/min/mg soluble protein across this CO_2 treatment range. Leaf soluble protein was nearly constant at about 2.4 g/m^2 with 55% being rubisco protein. Specific leaf weight increased across the 160 to 990 μ mol/mol CO_2 concentration range, so that the rubisco activity on a leaf dry weight basis decreased.

Campbell *et al.* (1990) showed that photosynthetic capacity of soybean canopies grown at 330 and 660 m mol/mol CO_2 were similar for short exposures at concentrations below 500 m mol/mol, but above this concentration, canopies grown at 660 m mol/mol had a slightly higher photosynthetic capacity. Thus, soybean did not lose photosynthetic capacity as did some other plant species. In other studies, however, both photosynthetic rate and rubisco activity of soybean declined during long-term CO_2 enrichment .

Respiration

Whole-canopy pre-dawn respiration rates at 40 to 60 DAP were 2-3 and 4-5 m mol/m²/s for soybean grown at CO_2 levels of 320 and 640 m mol/m², respectively, and a constant day/night temperature of 25°C (Jones *et al.*, 1985c). After cross-switching two of the treatment chambers at DAP 52, the subsequent canopy photosynthetic rates and respiration rates quickly adjusted to their new CO_2 exposure conditions.

Table. Average asymptotic maximum photosynthetic rate (Y_{max}) with respect to y-intercept parameter (Y_i), apparent Michaelis-Menten constant for CO_2 (K_m, and CO_2 compensation point (Γ_c.) for leaves grown at two CO_2 treatments and subjected to different short-term CO_2 levels.

Growth CO_2 treatment	*Y_{max} μ mol/m²/s*	*K_m μ mol/mol*	*Y_i μ mol/m²/s*	*μ $_c$ μ mol/mol*
330	51.8	359	-7.8	63
660	126.6	1 133	-4,6	42

Means of Y_{max} and Y_i were significantly different, p = 0.05, by a t-test. Thus, pre-dawn respiration rates were closely connected to the previous CO_2 fixation rates.

Partitioning

Growth of plants under elevated CO_2 results in changes in partitioning of photoassimilates to various plant organs over time. In soybean, elevated CO_2 generally promoted greater carbon (dry matter) partitioning to the supporting structure (stems, petioles and roots) than to the leaf laminae during vegetative stages of growth. During reproductive stages, there tended to be lower relative partitioning to reproductive growth (pods) by plants under elevated CO_2.

Growth Rates: During the linear phase of vegetative growth after full ground cover is reached, the growth rates of plants exposed to a range of CO_2 concentrations varied from 5.0 to 20.7 g/m²/d for exposures from 160 to 990 μ mol/mol.

The total final dry weight ranged from 12.88 to 39.12 g/plant, and final seed weight ranged from 5.77 to 17.85 g/plant for CO_2 treatments ranging from 160 to 660 μ mol/mol.

Carbohydrates

Soybean accumulates non-structural carbohydrates, particularly starch, under CO_2 enrichment. Allen *et al.* grew soybean under CO_2 treatments of 330, 450, 600 and 800 μ mol/mol. Sucrose, reducing sugars, and total soluble sugars of leaves remained somewhat constant throughout the day, but starch increased steadily at a rate of about 6 g/kg dry matter/hour. Average total soluble sugars increased from

24 to 36 g/kg dry weight and starch increased from 85 to 204 g/kg dry weight across the range of 330 to 800 μ mol/mol. Elevated non-structural carbohydrates in CO_2-enriched soybean plants were confirmed by Baker *et al.* and Allen *et al.* The concentrations also varied across the life cycle of the plants.

Table: *Soybean plant components as a percentage of total dry matter grown at subambient and superambient concentrations of CO_2 in 1984.*

Component	***CO_2 concentration, μ mol/mol***					
	160	***220***	***280***	***330***	***660***	***990***
	13 DAP [1] (V2 Stage)					
Root, %	11.3	9.8	9.3	12.4	9.8	10.5
Cotyledon, %	19.5	15.6	13.9	13.8	11.2	9.4
Stem, %	22.8	25.0	25.8	25.5	26.3	26.7
Leaf, %	46.1	49.5	50.7	48.2	52.5	53.2
	34 DAP (V8 Stage)					
Root, %	6.4	7.1	6.9	7.6	8.9	7.9
Stem, %	23.4	24.4	27.6	25.9	28.2	29.3
Petiole, %	12.2	13.1	14.6	14.4	15.7	15.7
Leaf, %	57.6	55.3	50.8	52.2	47.2	46.9
	66 DAP (R5 Stage)					
Stem, %	16.5	19.6	25.0	21.7	25.2	26.8
Petiole, %	9.9	11.3	12.9	12.7	14.1	13.6
Leaf, %	35.3	35.3	35.4	32.2	31.5	29.8
Pod, %	38.2	33.6	26.5	33.3	28.9	29.6
	94 DAP (R7 Stage)					
Stem, %	10.2	12.8	14.9	14.3	16.3	19.5
Petiole, %	5.5	6.2	6.0	6.7	7.0	6.5
Leaf, %	17.4	16.3	14.9	14.4	11.9	10.9
Pod, %	66.7	64.7	64.1	64.6	64.7	64.0

[1] Days after planting.

Nitrogen

Nitrogen content decreased from 50 to 37 g/kg dry weight over the range of 330 to 800 m mol/mol. When the nitrogen content was adjusted to remove the effect of total non-structural carbohydrate, the relative changes were smaller, 55 to 48 g N/kg dry matter, across the 330 to 800 μ mol/mol range.

Yield

Soybean seed yield was always increased by elevated CO_2 Allen *et al.* summarized the photosynthetic, biomass and seed yield responses of several experiments with the equation rectangular hyperbola model using data normalized to responses obtained at 330 μ mol/mol. The values of K_m , Y_{max} and Y_i parameters for relative photosynthetic rates were 279 m mol/mol, 3.08 and -0.68, respectively, for relative biomass yield were 182 m mol/mol, 3.02 and -0.91, respectively, and for relative seed yield were 141 m mol/mol, 2.55 and -0.76, respectively.

This model was used to project yields across several ranges of atmospheric CO_2 concentration increases. For a doubling of CO_2 this model predicted a 32.2% increase in soybean grain yield and a 42.7% increase in biomass. The ratio of these two numbers, 1.322/1.427 = 0.926, gives the fraction of the harvest index expected under doubled CO_2 in comparison with ambient CO_2.

Table: *Percentage increases of soybean midday photosynthetic rates, biomass yield, and seed yield predicted across selected carbon dioxide concentration [CO_2] ranges associated with relevant benchmark points in time. Adapted from Allen* et al.

Period of time (years)	*[CO_2]-Midday*		*Biomass photosynthesis*	*Seed yield*	*Biomass yield*
	Initial	*Final*			
	(Nmd/mol)		*(% increase over initial [CO_2])*		
IA-1700 [1]	200	270	38	33	24
1700-1973	270	330	19	16	12
1973-2073? [2]	330	660	50	41	31

1. IA, the Ice Age about 13 000 to 30 000 years before present. The atmospheric CO_2 concentrations that prevailed during the last Ice Age, and from the end of the glacial melt until pre-pioneer/pre-industrial revolution times, were 200 and 270 μ mol/mol, respectively.
2. The first world energy ‘crisis’ occurred in 1973 when the CO_2 concentration was 330 μ mol/mol. This CO_2 concentration is used as the basis for many CO_2 doubling studies. The CO_2 concentration is expected to double sometime within the 21st century.

Carbon Dioxide and Temperature

The CO_2 fertilization effect appears to be enhanced under elevated temperatures, at least up to a point. Idso *et al.* and Kimball *et al.* showed that the growth modification factor (or biomass growth modification ratio) due to a 300 μ mol/mol enrichment was 0.08 per °C (average daily temperature) across the range of 12 to 34°C. However, for soybean, Allen calculated a biomass growth modification ratio response to temperature of -0.031 per °C for seed biomass yield and -0.026 per °C for total biomass accumulation. Elevated temperatures tended to shorten the grain-filling period of this crop.

Soybean seed yield tended to decrease slightly with temperature over the day/night range of 26/19 to 36/29°C. The number of seed per plant increased slightly with increase of both CO_2 and temperature. Mass per seed decreased sharply with increasing temperature. Although CO_2 enrichment resulted in increased seed yield and above-ground biomass, harvest index was decreased with both CO_2 and temperature. The data of table show no tendency for the growth modification factor to increase with temperature for either seed yield or biomass accumulation.

Subsequent experiments with soybean substantiate these data although it was found that seed yields dropped sharply for day/night temperatures of 40/30°C and above, and biomass yields were maintained up to 44/34°C before rapidly failing.

The sensitivity of the growth modification factor data of Idso *et al.* and Kimball *et al.* may have been affected by other factors at Phoenix, Arizona. Apparently, the growth vs. temperature data were obtained throughout the year, and the findings may have been impacted by factors such as total solar radiation, photoperiod, stages of development, or other conditions caused by the changing seasons. Of course, these other environmental factors are all part of the complex of plant responses to climatic conditions. However, a better test of temperature effects alone would be CO_2 enrichment throughout the season or life cycle of plants under natural conditions, but with consistent temperature differences (cooler and warmer).

Table: *Seed yield, components of yield, total above-ground biomass and harvest index of soybean grown at two CO_2 concentrations and three temperatures in 1987*

CO_2 conc. (μ mol/mol)	Day/night temperature (°C)	Grain yield (g/plant)	Seed/plant (no./plant)	Seed mass (mg/seed)	Above-ground biomass (g/plant)	Harvest index
330	26/19	9.0	44.7	202	17.1	0.53
330	31/24	10.1	52.1	195	19.8	0.51
330	36/29	10.1	58.9	172	22.2	0.45
660	26/19	13.1	58.8	223	26.6	0.49
660	31/24	12.5	63.2	198	27.6	0.45
660	36/29	11.6	70.1	165	26.5	0.44

F-values					
CO_2 conc.	12.3**	11.4**	2.5*	NA	NA
Temperature	0.0 NS	8.4**	106.2**	NA	NA
CO_2 x Temperature	2.0 NS	0.1 NS	11.2**	NA	NA

*, ** Significant at the 0.05 and 0.01 probability levels, respectively.

NS = not significant; NA = *not* available.

Evapotranspiration and Water-use Efficiency

The direct effect of increasing temperatures across the range of 28 to 35°C appears to increase transpiration rate about 4 to 5% per °C, based on both experimental and modelling studies. This is in close agreement with the rise in saturation vapour pressure of about 6% per °C. Allen *et al.* showed that the increase in WUE of soybean contributed by increased photosynthesis under elevated CO_2 was much greater than the increase in WUE contributed by decreased transpiration. The fractional increase in WUE attributable to increased photosynthesis and decreased transpiration were about 0.8 and 0.2, respectively.

Rice

Leaf and Canopy Photosynthesis: Leaf photosynthetic rates, in high light, of cv. IR72 rice, determined in 1992 (RICE VI experiment) for plants grown at three day/night temperature regimes, 32/23, 35/26 and 38/29°C (but measured at ambient temperature), averaged 18.8 and 30.4 μ mol $CO_2/m^2/s$ for 330 and 660 μ mol/mol CO_2 treatments, respectively.

This increase caused by elevated CO_2 is about 60%. Leaf A/C_i curves obtained from both CO_2 treatments at the growth temperature of 35/26°C. There were essentially no differences in A between the two CO_2 treatments at each specific C. level. Similar pairs of curves were obtained for the other temperature treatments. Each pair of leaf A/C_i curves were similar within each temperature regime. Thus, the leaf A/C_i curves gave no indication of a change in photo-synthetic capacity. Extrapolations of the response curves to zero values of A gave an intercept (leaf CO_2 compensation point) of about 60 μ mol/mol CO_2.

Following canopy closure, the rice canopy net photosynthetic rate (Pn) vs. photo-synthetic photon flux density (PPFD) responses were linear and did not approach light saturation in any of the experiments, probably because of the erect leaf orientation of cv. IR30 and high plant populations. Canopy Pn vs. PPFD at 60 DAP for the six CO_2 treatments of the RICE II experiment (Baker and Allen, 1993b; Allen *et al.,* 1995) gave values of about 34, 50, 60, 80, 85 and 90 μ mol $CO_2/m^2/s$ at high light (1600 μ mol photons/m^2/s) for treatments of 160, 250, 330, 500, 660 and 900 μ mol/mol.

Linear canopy Pn responses to PPFD on day 60 of the RICE IV experiment were similar among all temperature treatments (25/18,

28/21, 31/24, 34/27 and 37/30°C) that were exposed to 660 μ mol/mol CO_2 responses of the 330 μ mol/mol treatment were about 25% less. Other studies of canopy Pn have shown only small differences across wide ranges of temperature; e.g., cotton and soybean.

There may be two reasons for the lack of a clear canopy photosynthetic response to air temperature across the 25 to 37°C range. Evaporative cooling may lower foliage temperature below air temperature increasingly with increasing air temperature and increasing vapour pressure deficit. The summed response of the photosynthetic rates of leaves to temperature at all exposures of light may broaden the temperature response for the whole canopy photosynthetic rates.

Baker and Allen fitted whole-day canopy Pn data to a rectangular hyperbola of the form of equation for DAP 61 of the RICE II experiment. The parameters were 70.8 μ mol/mol, 3.96 and -2.21, for K_m, Y_{max} and Y_i, respectively, with a relative Pn response ceiling of 1.75 at infinite [C]. The calculated percentage increase in response of whole-day canopy Pn at 660 vs. 330 μ mol/mol was 36%.

Photosynthetic Acclimation to CO_2 at the Canopy Level

To test for acclimation of canopy photosynthetic capacity, Baker *et al.* used rice that was grown at 160, 250, 330, 500, 660 and 900 μ mol/mol CO_2. For half-day periods (mornings) on 62,63 and 64 DAP, common CO_2 setpoints of 160, 330 or 660 m mol/mol were imposed on each chamber on those respective days.

Within each of these short-term CO_2 exposure comparisons, Baker *et al.* and Baker and Allen (1993a) showed that the short-term canopy net photosynthetic rate decreased with increasing long-term CO_2 treatment.

The relative effects of the short-term CO_2 exposure were greatest for the lowest short-term CO_2 concentration. For example, when compared at a common short-term CO_2 exposure of 160 μ mol/mol, the canopy photosynthetic rate of the chamber containing the 900 μ mol/mol long-term CO_2 growth treatment was only about one-third of that of the 160 μ mol/mol long-term CO_2 growth treatment.

However, a large part of this apparent acclimation effect may be attributed to the greater respiration rates of the plants that had been grown under elevated CO_2.

Rubisco protein percentage was used as evidence of leaf acclimation to a wide range of CO_2 concentrations for 34 DAP soybean and 75 DAP rice. For rice, rubisco activity expressed on a leaf area basis decreased by 66% across the 160 to 900 μ mol/mol long-term CO_2 treatments. A major cause of this decline in rubisco activity was a 32% decrease in the amount of rubisco protein relative to other soluble protein.

Although Rowland-Bamford *et al.* reported decreases in both amount and activity of rubisco with increasing CO_2 in rice plants grown across the range of 160 to 900 μ mol/mol, the leaf A/C. response curves gave no indication of a downward acclimation of photosynthetic capacity across the smaller range of 330 to 660 μ mol/mol CO_2.

One possibility would be that rubisco is not as limiting for photosynthesis under CO_2 enrichment as would be expected. More work is needed, under a range of CO_2 treatments, to explore the interaction effects of sink capacity, nitrogen nutrition, and other internal CO_2-fixation processes on photosynthetic behaviour and crop yield.

***Table:** For soybean, leaf blade soluble protein expressed on a leaf blade area basis and percentage rubisco protein expressed on a leaf blade soluble protein basis for 34-day-old soybean plants grown under a wide range of CO_2 concentrations. For rice, leaf nitrogen content expressed on a leaf area basis and percentage rubisco protein expressed on a leaf soluble protein basis for 75-day-old rice plants grown under a wide range of CO_2 concentrations.*

Soybean			*Rice*		
CO_2 growth concentration (μ mol/mol)	*Leaf soluble protein. (g/m²)*	*Rubisco protein (%)*	*CO_2 growth concentration (μ mol/mol)*	*Leaf nitrogen protein (μ mol/m²)*	*Rubisco protein (%)*
160	2.5	56	160	95	62
220	3.2	54	250	90	59
280	2.6	-	330	81	54
330	2.3	57	500	62	49
660	2.3	54	660	78	43
990	2.3	55	900	64	42

Development Stages

In rice, two distinctly different rates of leaf appearance for vegetative and reproductive phases of growth were found, as has been observed many times before. Leaf appearance rates (leaves per day) are about twice as great in the vegetative phase as in the reproductive phase.

In the 'late' planted RICE II experiment, the number of mainstem leaves at panicle initiation and the final number of mainstem leaves decreased across CO_2 treatments of 160 to 500 μ mol/mol and remained similar from 500 to 900 μ mol/mol. Panicle initiation and boot stage occurred about 12 days earlier in the superambient CO_2 treatments compared with the 160 μ mol/mol treatment. Therefore, plant developmental rate was clearly accelerated with increasing CO_2 up to about 500 μ mol/mol.

Temperature effect data were assembled from treatments ranging from 25/18/21 to 40/33/37°C (day/night/paddy water temperatures) in RICE I to RICE V experiments. Air temperature greatly influenced leaf appearance rate, developmental rate, and total growth duration, whereas, across the limited range of 330 to 660 μ mol/mol the effects were comparatively small. No consistent differences in phyllochron interval (days per leaf appearance) between 330 and 660 μ mol/mol were observed, whereas phyllochron interval increased with increasing temperature across the range of 25/18/21 to 40/33/37°C, especially for the reproductive phase. In RICE III, IV and V experiments, anthesis ranged from 0 to 6 days earlier in the 660 vs. 330 μ mol/mol treatments. Also, the number of days to anthesis was shortened by approximately 10 days across the temperature treatment range from 25/18/21 to 34/27/31°C for the RICE IV experiment.

Canopy Dark Respiration Rates

Plant respiration rates may be decreased by both short- and long-term exposure to high CO_2 concentrations; however Baker *et al.* found that nighttime canopy dark respiration rates [R_d, μ mol $(CO_2)/m^2$ (ground area)/s] increased for rice exposed to daytime CO_2 ranging from 160 to 900 μ mol/mol. Similar to photosynthetic rates, R_d increased with CO_2 exposure from 160 to 500 μ mol/mol but levelled off somewhat across the 500 to 900 μ mol/mol range. The R_d of the ambient and superambient CO_2 treatments reached a broad maximum around 30 to 50 DAP whereas the broad maximum of the subambient treatments occurred later around 50 to 70 DAP. The maximum values of R_d were about 6, 8, and 9 μ mol/m²/s for the 160, 250 and 330 μ mol/mol CO_2 treatments and about 11 to 12 μ mol/m²/s for the three superambient CO_2 treatments.

Specific respiration rate [R_{dw} , m mol (CO_2)/s/kg (total above-ground dry matter)] decreased exponentially with DAP at all CO_2

exposures and was higher in the subambient (160 and 250 μ mol/mol) than in the ambient (330 μ mol/mol) and superambient (500, 660 and 900 *m* mol/mol treatments). At each CO_2 exposure level, the patterns of R_{dw} with DAP were very similar to the patterns of plant tissue nitrogen concentration with DAP. Furthermore, R_{dw} was linearly related to total above-ground plant tissue nitrogen concentration [mg (N)/g (DW)] across the range of CO_2 exposures and the six dates of plant sampling (R_{dw} = -13.0 + 0.952 * [N]; r2=0.9, P=0.01). Baker *et al.* concluded that the CO_2 treatments affected R_{dw} mainly by altering the protein content of the plant tissue. Another explanation is that elevated CO_2 increased the amount of structural and non-structural carbohydrates in the plant tissues, so that a larger proportion of dry matter was sequestered in non-protein materials. This study showed no respiration acclimation to long-term CO_2 enrichment (indirect acclimation) of rice that could not be explained by nitrogen concentration of the plant tissues.

Evapotranspiration and Water-use Efficiency

Both plant transpiration and direct evaporation from the floodwater surface contribute to water use (evapotranspiration, ET) of rice growing in the SPAR chambers. Diurnal trends of ET followed diurnal patterns of solar irradiance. After canopy closure, maximum ET rates and total daytime water losses were about 35 and 30% greater, respectively, from rice grown at 160 μ mol/mol compared to rice grown at 900 μ mol/mol CO_2. The ET rates were similar when all chambers were exposed for one-half day to the same CO_2 concentration, which demonstrates the effect of the exposure level of CO_2 on stomatal control of transpiration. Temperature also has a large effect on diurnal ET rates and daytime water use, mediated primarily through vapour pressure deficit of the air. Solar irradiance also has a large effect, probably through both the energy inputs to the canopy and through the stomatal opening response to light (directly or indirectly). Midday maximum ET rates were about 75 and 35% higher for rice grown at 40/33/37°C and 34/27/31°C, respectively, compared to rice grown at 28/21/25°C.

Water-use Efficiency: Effects of CO_2

Daytime totals of CO_2 uptake, ET, and calculations of WUE vs. CO_2 treatment are shown in table. Stomatal conductance decreases with increasing CO_2 concentration which can cause a reduction of both leaf and whole canopy transpiration. However, CO_2 enrichment may also increase canopy leaf surface area for transpiration, thereby offsetting some of the water savings. The leaf area index of the RICE II experiment ranged from 7.6 to 10.8 across the CO_2 treatments from 160 to 900 μ mol/mol. Calculations of WUE increased with increasing CO_2 due to the decline in ET across this CO_2 range and the increase in Pn with CO_2 up to the 500 μ mol/mol treatment.

Table: *Comparison of total daytime responses of rice canopies grown season-long in various CO_2 concentration treatment regimes. Daytime dry bulb air temperature/nighttime dry bulb air temperature/paddy water temperature: dewpoint temperature treatments are given in the second column.*

CO_2	Temperature [1]	Total daytime photons	Total daytime CO_2 uptake	Total daytime H_2O loss	Water-use efficiency [2]
μ mol/mol	°C	mol(Photons)/m²	mol(CO_2)/m²	mol(H_2O)/m²	mmol/mol
Effects of carbon dioxide: RICE II, 23 August 1987, 61 DAP					
160	31/31/27:18	39.4	0.74	608.5	1.22
250	31/31/27:18		1.13	643.4	1.76
330	31/31/27:18		1.34	611.0	2.19
500	31/31/27:18		1.80	553.6	3.25
660	31/31/27:18		1.79	536.2	3.34
900	31/31/27:18		1.83	469.5	3.90
Effects of temperature: RICE IV, 10 September 1989, 58 DAP					
330	28/21/25:12.0	26.4	0.87	612.7	1.42
660	25/18/21:10.5		0.83	359.3	2.30
660	28/21/25:12.0		1.13	475.8	2.37
660	31/24/28:13.5		0.83	491.3	1.69
660	34/27/31:15.0		0.89	736.4	1.20
660	37/30/34:16.5		0.97	909.4	1.06

1. Daytime dry bulb air temperature/nighttime dry bulb air temperature/paddy water temperature: dewpoint temperature.
2. mmol CO_2/mol (H_2O).

Water-use Efficiency: Effects of Temperature

In the 28°C air temperature treatment, CO_2 enrichment from 330 to 660 *m* mol/mol resulted in 30% increase in daytime total CO_2 uptake and 22% decrease in ET with a concomitant increase of WUE from 1.42 to 2.37 m mol (CO_2)/μmol (H_2O), an increase of 67%. Daytime ET was more than doubled from the 25 to the 37°C daytime temperature treatment. WUE declined from the 28 to the 37°C daytime treatment due to the sharp increase in ET and the relatively stable Pn across this temperature range. If WUE were based on grain yield, it would decrease drastically with increasing temperature not only because of increasing ET, but also because of sharply decreasing seed production.

Rising atmospheric CO_2 is likely to benefit rice production by increasing photosynthesis, growth and grain yield while reducing water use and increasing WUE. In warm areas of the world, however, possible future global warming may result in substantial yield decreases because of the sensitivity of flowering and seed set to high temperatures and the possibility of water shortages that may result from increased evapotranspiration.

Summary of Comprehensive Reviews

Several recent symposia proceedings and reviews leave little doubt that crop plants can respond well to elevated CO_2. Poorter compiled information from 156 plant species and found that doubling CO_2 provided an average growth increase of 37%. The distribution of weight ratios of CO_2-enriched and control plants is shown in As a group, C_3 herbaceous crop plants responded more than wild herbaceous species (58 vs. 35%). Furthermore, the fast growing wild species responded more strongly than slow-growing wild species (54 vs. 23%).

Poorter (1993) imposed two restrictions on his compilations that may have led to larger than expected responses to elevated CO_2. Firstly, plants grown in competition were not included. Secondly, only vegetative stages of plants were compared since compiled data were selected prior to flowering.

A number of studies have shown that vegetative growth responses may be greater than reproductive responses. Therefore, the compiled data of Poorter may give an impression of greater response than would be observed throughout the life cycle. Secondly, crops in field conditions usually are grown in dense populations where they compete for space and light. Under more realistic field conditions, crop plants are likely to respond as a community rather than individual plants, wherein light (solar radiation) becomes a limiting factor for growth. Under these conditions, elevated CO_2 cannot promote horizontal expansion and greater light capture. Although the actual field responses may be less, the CO_2 fertilization effect is clearly well-established.

The CO_2 fertilization effect for forest species has also become firmly established. Wullschleger *et al.* estimated the biotic growth factor for 58 controller-exposure studies of forest tree species which included 398 observations. Their frequency distribution of relative growth response of trees grown at elevated CO_2 vs. ambient CO_2. They also found a mean response ratio of 1.32. Of the 398 observations, 51 showed a relative growth response less than 1, and 31 showed responses greater than 2. However, under competitive conditions, tree response may be much less. Kimball *et al.* discussed the data of Idso *et al.* (1987) which related the growth modification factor (or biomass growth modification ratio) caused by a 300 μ mol/mol increase in CO_2 concentration above ambient.

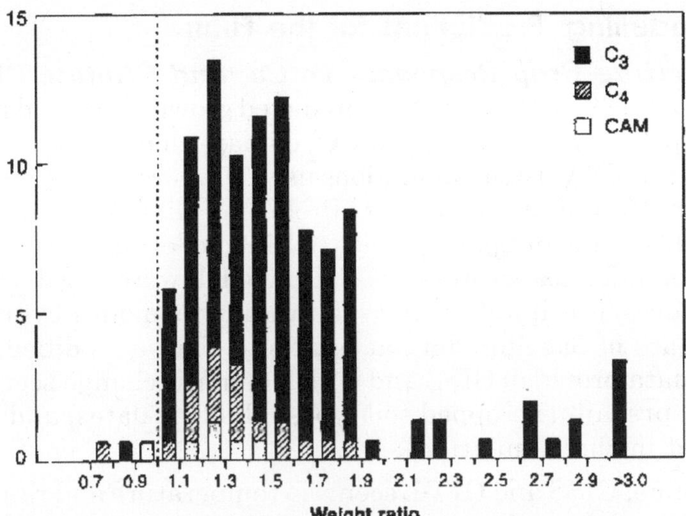

Figure: *Distribution of the biomass growth modification ratio (weight ratio) of CO_2-enriched plants (600 to 720 μ mol/mol) in comparison with control treatments (300 to 360 μ mol/mol). The bar graph was created from averages of all the weight ratios of 156 species selected from the literature.*

Their data are reproduced. Under Phoenix, Arizona, conditions, the growth modification factor across a mean temperature range of 12 to 34 °C was rather linear. Thus, the *vegetative* response to CO_2 should be enhanced at increased temperatures. As discussed before, this growth modification factor may not apply for reproductive growth responses of crops (e.g., rice).

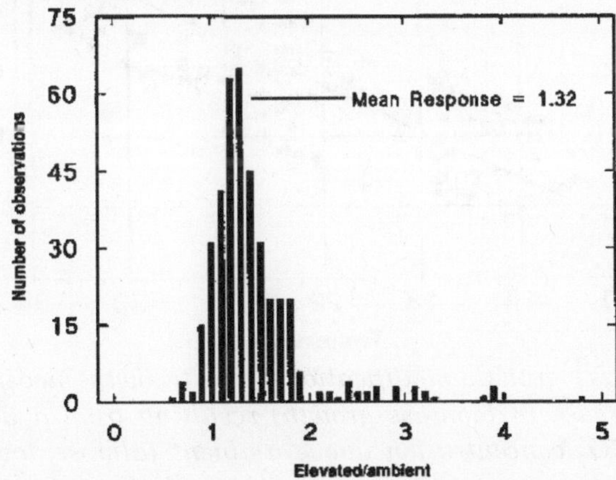

Figure: *Frequency distribution of the log-transformed biomass growth modification ratio (log elevated/ambient CO_2 exposure total dry mass ratios) of 73 tree species.*

Crop Modelling: Predictions for the Future

Modelling Crop Responses To Co_2 and Climate Changes : Peart *et al.* (1989) and Curry *et al.* predicted growth and yield responses of soybean and maize to doubled-CO_2 climate change scenarios of the southeastern USA. Their simulations used 30 years of baseline weather data from 19 sites in 11 states. Predicted climate changes of the Goddard Institute for Space Studies (GISS) model and the Geophysical Fluid Dynamics Laboratory (GFDL) model were used to change temperatures, precipitation and solar radiation, month-by-month, for the 30 years of baseline data at each site. These modified baseline weather data provided GISS and GFDL climatic change scenarios. At each site, prevailing cropped soil types, planting dates, and cultivars were used in the simulations.

Baseline, GISS and GFDL scenario temperature and rainfall data that were used by Peart *et al.* (1989) were condensed for two representative locations; Columbia, South Carolina, and Memphis, Tennessee. Crop yield responses to climate change were simulated under four conditions: with or without direct CO_2 fertilization effects, and under rainfed or optimum irrigation culture.

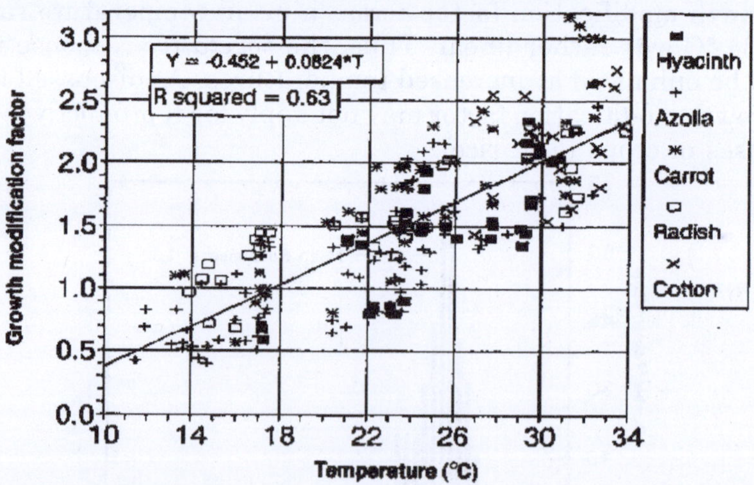

Figure: *Biomass growth modification ratio (growth modification factor or relative increase in biomass growth) resulting from a 300 μ mol/mol increase in CO_2 concentration above ambient (almost doubled) versus mean daily air temperature for the plants indicated in the legend.*

The Penman-Monteith equation, which contains a term for canopy conductance, was used to compute the effects of elevated CO_2 on canopy transpiration. Simulations were run for doubled-CO_2 conditions with a crop photosynthetic enhancement factor of 1.35 for soybean (a C_3 plant) and 1.10 for maize (a C_4 plant). No simulations were run with CO_2 fertilization effects only (without climate change effects).

Simulated soybean seed yields were averaged over 30 years and 19 sites. Under rainfed conditions with climate change effects only, simulated soybean yields under the GFDL scenario were reduced 71 % compared to the baseline climate, whereas the average yields were reduced only 23% under the GISS scenario. Yields under the GFDL scenario were severely impacted because of the rainfall reductions predicted by this model. Under optimum irrigation conditions, average yields under both the GISS and GFDL scenarios were reduced 18 to 19%. However, in spite of higher temperatures, the irrigated yields under the GISS and GFDL scenarios were about 25% greater than the baseline climate scenario without irrigation.

Table: *Summary of 30-year mean (1951-1980) baseline (BASE) weather data of two locations with GISS and GFDL scenarios.*

	BASE			***GISS***			***GFDL***		
	Prec. (mm)	***T_{max} (°C)***	***T_{min} (°C)***	***Prec. (mm)***	***T_{max} (°C)***	***T_{min} (°C)***	***Prec. (mm)***	***T_{max} (°C)***	***T_{min} (°C)***
Columbia. SC									
APR-SEP; Jul	684	33.3	21.2	817	35.2	23.1	471	38.2	26.1
OCT-MAR; Jan	561	13.5	0.7	602	15.8	3.0	571	15.6	2.8
TOTAL; Mean	1245	24.1	10.7	1419	26.6	13.2	1042	27.3	13.9
Memphis. TN									
APR-SEP; Jul	656	33.1	22.6	748	35.4	24.9	549	36.0	25.5
OCT-MAR; Jan	654	9.0	-0.6	565	12.0	2.4	758	11.3	1.7
TOTAL; Mean	1 310	22.0	11.1	1 313	25.4	14.5	1307	24.8	13.9

When CO_2 fertilization effects were included with climate change effects, simulated average yields of the GISS climate scenario were increased 11% under rainfed conditions, whereas yields under the GFDL climate scenario were still decreased (-52%). Under optimum irrigation with CO_2 fertilization effects, yields under both GISS and GFDL scenarios were increased 13 to 14%.

Yields of maize were simulated for doubled-CO_2 fertilization effects at only four weather station locations (Charlotte, North Carolina; Macon, Georgia; Memphis, Tennessee; Meridian, Mississippi). For climatic change effects alone, predicted maize yields declined only 6% in the GISS scenario, but declined 73% in the GFDL scenario.

Although irrigation increased predicted crop yields, the GISS and GFDL climate scenarios gave yield decreases of 18 and 27%, respectively, with respect to the irrigated baseline weather. The yield reduction of the GFDL scenario with respect to the GISS scenario was 10%, attributable to slightly higher tempera-tures.

Including the CO_2 fertilization effects with climatic change scenarios had little effect on the predicted yields of maize because it is a C_4 plant.

In the Great Lakes and Corn Belt Area, simulations by Ritchie *et al.* showed that higher temperature would have the greatest effect of the climate change factors on predicted yields of soybean and maize, mainly through decreases in the crop life cycle.

For the southern part of this region, yield reductions were greater for the GFDL scenario. Predicted yields increased for the northernmost stations because temperatures and growing season duration became more favourable. Average irrigation water requirements in this region increased about 90% for the GISS and GFDL scenarios.

Rosenzweig modelled maize and wheat yields in the Great Plains under GISS and GFDL scenarios. Yields decreased in the Southern Great Plains because higher temperatures shortened the life cycle of the crops. Where precipitation was predicted to decrease, irrigation requirements increased. In a separate modelling study, Allen and Gichuki predicted a 15% increased requirement for irrigation for this region, with greater requirements for alfalfa because its growing season was increased, and lower requirements for maize and winter wheat because their growing seasons were decreased. The CO_2 fertilization effect offset the adverse effects of climate change at some locations of Ritchie *et al.* and Rosenzweig.

Table: *Doubled CO_2 soybean yield simulations (SOYGRO) for the southeastern USA*

BASE	***GISS Model***		***GFDL Model***		***Model***
Yield (kg/ha)	***Yield (kg/ha)***	***% Diff.***	***Yield (kg/ha)***	***% Diff.***	***% Diff.***
Climate Change Effects Only, Rainfed					
2497	1 929	-23	733	-71	-62
Climate Change Effects Only, Irrigated					
3 837	3 158	-18	3092	-19	- 2
CO_2 Fertilization plus Climate Change Effects, Rainfed					
2497	2780	+11	1 206	-52	-57
CO_2 Fertilization plus Climate Change Effects. Irrigated					
3837	4350	+13	4393	+14	+1

Adapted from Peart *et al.* (1989).

Table: *Doubled CO_2 maize yield simulations (CERES-Maize) for Charlotte NC, Macon GA, Meridian MS, and Memphis TN*

BASE	***GISS Model***		***GFDL Model***		***Model***
Yield (kg/ha)	***Yield (kg/ha)***	***% Diff.***	***Yield (kg/ha)***	***% Diff.***	***% Diff.***
Climate Change Effects Only, Rainfed					
8468	7926	-6	2289	-73	-71
Climate Change Effects Only. Irrigated					
13,899	11 455	-18	10257	-26	-10
CO_2 Fertilization plus Climate Change Effects. Rainfed					
8577	8 136	-5	2224	-74	-73
CO_2 Fertilization plus Climate Change Effects. Irrigated					
14052	11 545	-18	10363	-26	-10

Dudek (1989) predicted productivity changes of several Californian vegetable, fruit and nut crops in response to GISS and GFDL scenarios for doubled CO_2. Without CO_2 fertilization effects, statewide average yield changes, depending on the crop, would be -8 to -34% for the GISS scenario and -6 to -31 % for the GFDL scenario.

With CO_2 fertilization effects, predicted statewide yield changes ranged from about +17 to -12% for the GISS scenario and about +21 to -8% for the GFDL scenario. Dudek (1989) used a California Agriculture and Resource Model (CARM) to further predict economic and market impact of the productivity changes. Production declined, in general, under the scenarios of climate change without CO_2 fertilization effects; however, commodity prices generally increased. Under the CO_2 fertilization plus climate change effects scenarios, predicted impacts on commodity prices were much less.

The effect of temperature on the phenology of crop plants plays a critical role. One crucial need is for more detailed research on responses of plants to temperature, and temperature x CO_2 interactions, as inputs to crop models. More modelling studies are also needed on different planting dates as an adaptive strategy. Cultivars need to be designed for future climatic conditions.

Factors that should be considered are: extension of the grain-filling period and perhaps shortening the duration of vegetative growth (which would also improve harvest index); ability to flower and set seed at higher temperatures; photoperiod and thermoperiod adaptive interactions; selection for positive photosynthetic acclimation where negative photosynthetic acclimation has been observed; and capability of utilizing photoassimilates (storage carbohydrates) more effectively. These factors need to be integrated into whole-plant physiology under real-world conditions.

Adaptations And Evapotranspiration Requirements

The simulated crop yield responses to climatic changes provided by Peart *et al.* and Curry *et al.* manifest two main points: (a) the serious adverse impact of inadequate rainfall scenarios on crop production coupled with rising temperature scenarios, and (b) the importance of beneficial CO_2 fertilization effects in the face of elevated temperatures. However, the climate change for an effective doubling of CO_2 may occur at CO_2 concentrations less than those used in this simulation, if radiatively active trace gases other than CO_2 play a large role in the greenhouse effect. All of the simulations assumed that climatic changes would occur simultaneously with increasing concentrations of CO_2 and other trace greenhouse-effect gases.

If global warming lags the increases of atmospheric CO_2, then some beneficial effects of CO_2 fertilization are likely to occur before the full impact of climate change is manifested. However, Broecker and others caution that climate changes have not always been gradual during interglacial periods of the Pleistocene Era. There is clear evidence of relatively rapid climatic oscillations in the northern hemisphere during the previous interglacial period 110000 to 140000 years before present based on Greenland Ice-core Project (GRIP) records. These oscillations produced cold periods that were as severe as the proceeding glacial period.

Much of the reduction in soybean yields reported by Peart *et al.* and Curry *et al.* was caused by decreases in the length of the grain-filling period under higher temperatures. Changes in management practices, such as changing planting dates or selection of other cultivars, may help to prevent some of the potential reductions in yield. In the future, plant breeders may need to adapt combinations of temperature tolerance and photoperiod responses into new germplasm. In situations where non-structural carbohydrates accumulate as a CO_2 fertilization effect response, new germplasm needs to be developed that can make better use of the photoassimilate source.

Irrigation is not likely to be a panacea for climate change. Predicted irrigation requirements for soybean in the southeastern USA were increased 33 and 134% under the GISS and GFDL scenarios, respectively, based on simulations of Peart *et al.* (1989) and Curry *et al.*. However, under the GFDL scenario, water resources would become scarce, and may not be readily available for crops. Some areas of the USA may have to adapt by irrigating less land area. Increasing temperatures and decreasing precipitation for the USA as predicted by the GFDL model would have a serious negative impact overall on agricultural productivity although producers in favourable regions may benefit from scarcity-mediated higher prices.

Assessing International Impacts of Rising Co_2 and Climate Change

Several assessments have been conducted on the impacts of rising CO_2 and global climatic changes on crop production patterns and economics responses within national or regional zones and in various countries around the world. Furthermore, progress has been made on predicting impacts of climate change using world trade models. The studies conducted by Rosenzweig and colleagues used existing crop models and climate change scenarios from three GCMs (GISS, GFDL and United Kingdom Meteorological Office Model (UKMO), Wilson and Mitchell, 1987) to predict changes in crop yields for a number of countries around the world. The doubled-CO_2 climate change scenarios (temperature, rainfall and evaporation changes) were based on the climate change potential expected from increases of all greenhouse-effect gases. These climate changes are expected to occur well before CO_2 concentration has actually doubled. Therefore, the CO_2 levels for fertilization effects were estimated to be 555 μ mol/mol rather than 660 μ mol/mol. Thus, the CO_2 fertilization effect photosynthetic ratios for four crops (soybean, wheat, rice and maize) were taken as 1.21, 1.17, 1.17 and 1.06, respectively.

The impact of climate change scenarios was more severe in the tropical latitudes than in the mid- or high-latitudes. For example, averaged over all three GCM scenarios, wheat production changes predicted in Brazil, India, China and Canada were -47, -43, -11 and -20%, respectively, without CO_2 effects, and -28, -13, +8 and +16% with CO_2 fertilization effects (calculated from data of Rosenzweig and Parry, 1993; Rosenzweig and Iglesias, 1994). Simulations of soybean yields throughout the USA were a part of this international study (Curry *et al.,* 1995). They found that aggregated yields were 2.42, 2.80, 2.37 and 1.31 Mg/ha for the BASE, GISS, GFDL and UKMO scenarios, respectively. Temperatures were about 4.0°C higher than BASE for the GISS and GFDL scenarios, but were about 5.2°C higher for the UKMO model.

Refinements and improvements in prediction methodology will continue, but these assessments provide the best currently available insights into the CO_2 fertilization and climate change effects on global crop productivity.

7
Climate Protection

Climate encompasses the statistics of temperature, humidity, atmospheric pressure, wind, precipitation, atmospheric particle count and other meteorological elemental measurements in a given region over long periods. Climate can be contrasted to weather, which is the present condition of these elements and their variations over shorter periods.

A region's climate is generated by the climate system, which has five components: atmosphere, hydrosphere, cryosphere, land surface, and biosphere.

The climate of a location is affected by its latitude, terrain, and altitude, as well as nearby water bodies and their currents. Climates can be classified according to the average and the typical ranges of different variables, most commonly temperature and precipitation. The most commonly used classification scheme was originally developed by Wladimir Köppen. The Thornthwaite system, in use since 1948, incorporates evapotranspiration along with temperature and precipitation information and is used in studying animal species diversity and potential effects of climate changes. The Bergeron and Spatial Synoptic Classification systems focus on the origin of air masses that define the climate of a region.

Paleoclimatology is the study of ancient climates. Since direct observations of climate are not available before the 19th century, paleoclimates are inferred from *proxy variables* that include non-biotic evidence such as sediments found in lake beds and ice cores, and biotic evidence such as tree rings and coral. Climate models are mathematical models of past, present and future climates. Climate change may occur over long and short timescales from a variety of factors; recent warming is discussed in global warming.

The difference between climate and weather is usefully summarized by the popular phrase "Climate is what you expect, weather is what you get." Over historical time spans there are a number of nearly constant variables that determine climate, including latitude, altitude, proportion of land to water, and proximity to oceans and mountains. These change only over periods of millions of years due to processes such as plate tectonics.

Other climate determinants are more dynamic: the thermohaline circulation of the ocean leads to a 5 °C (9 °F) warming of the northern Atlantic Ocean compared to other ocean basins. Other ocean currents redistribute heat between land and water on a more regional scale.

The density and type of vegetation coverage affects solar heat absorption, water retention, and rainfall on a regional level. Alterations in the quantity of atmospheric greenhouse gases determines the amount of solar energy retained by the planet, leading to global warming or global cooling. The variables which determine climate are numerous and the interactions complex, but there is general agreement that the broad outlines are understood, at least insofar as the determinants of historical climate change are concerned.

There are several ways to classify climates into similar regimes. Originally, climes were defined in Ancient Greece to describe the weather depending upon a location's latitude. Modern climate classification methods can be broadly divided into *genetic* methods, which focus on the causes of climate, and *empiric* methods, which focus on the effects of climate.

Examples of genetic classification include methods based on the relative frequency of different air mass types or locations within synoptic weather disturbances. Examples of empiric classifications include climate zones defined by plant hardiness, evapotranspiration, or more generally the Köppen climate classification which was originally designed to identify the climates associated with certain biomes. A common shortcoming of these classification schemes is that they produce distinct boundaries between the zones they define, rather than the gradual transition of climate properties more common in nature.

Bergeron and Spatial Synoptic

The simplest classification is that involving air masses. The Bergeron classification is the most widely accepted form of air mass classification. Air mass classification involves three letters. The first letter describes its moisture properties, with c used for continental air masses (dry) and μ for maritime air masses (moist).

The second letter describes the thermal characteristic of its source region: T for tropical, P for polar, A for Arctic or Antarctic, M for monsoon, E for equatorial, and S for superior air (dry air formed by significant downward motion in the atmosphere). The third letter is used to designate the stability of the atmosphere. If the air mass is colder than the ground below it, it is labelled k. If the air mass is warmer than the ground below it, it is labelled w. While air mass identification was originally used in weather forecasting during the 1950s, climatologists began to establish synoptic climatologies based on this idea in 1973.

Based upon the Bergeron classification scheme is the Spatial Synoptic Classification system (SSC). There are six categories within the SSC scheme: Dry Polar (similar to continental polar), Dry Moderate (similar to maritime superior), Dry Tropical (similar to continental tropical), Moist Polar (similar to maritime polar), Moist Moderate (a hybrid between maritime polar and maritime tropical), and Moist Tropical (similar to maritime tropical, maritime monsoon, or maritime equatorial).

Köppen

The Köppen classification depends on average monthly values of temperature and precipitation. The most commonly used form of the Köppen classification has five primary types labelled A through E. These primary types are A, tropical; B, dry; C, mild mid-latitude; D, cold mid-latitude; and E, polar. The five primary classifications can be further divided into secondary classifications such as rain forest, monsoon, tropical savanna, humid subtropical, humid continental, oceanic climate, Mediterranean climate, steppe, subarctic climate, tundra, polar ice cap, and desert.

Rain forests are characterized by high rainfall, with definitions setting minimum normal annual rainfall between 1,750 millimetres (69 in) and 2,000 millimetres (79 in). Mean monthly temperatures exceed 18 °C (64 °F) during all months of the year.

A monsoon is a seasonal prevailing wind which lasts for several months, ushering in a region's rainy season. Regions within North America, South America, Sub-Saharan Africa, Australia and East Asia are monsoon regimes.

A tropical savanna is a grassland biome located in semiarid to semi-humid climate regions of subtropical and tropical latitudes, with average temperatures remain at or above 18 °C (64 °F) year round and rainfall between 750 millimetres (30 in) and 1,270 millimetres (50 in) a year. They are widespread on Africa, and are found in India, the northern parts of South America, Malaysia, and Australia.

The humid subtropical climate zone where winter rainfall (and sometimes snowfall) is associated with large storms that the westerlies steer from west to east. Most summer rainfall occurs during thunderstorms and from occasional tropical cyclones. Humid subtropical climates lie on the east side continents, roughly between latitudes 20° and 40° degrees away from the equator.

A humid continental climate is marked by variable weather patterns and a large seasonal temperature variance. Places with more than three months of average daily temperatures above 10 °C (50 °F) and a coldest month temperature below "3 °C (27 °F) and which do not meet the criteria for an arid or semiarid climate, are classified as continental.

An oceanic climate is typically found along the west coasts at the middle latitudes of all the world's continents, and in southeastern Australia, and is accompanied by plentiful precipitation year round.

The Mediterranean climate regime resembles the climate of the lands in the Mediterranean Basin, parts of western North America, parts of Western and South Australia, in southwestern South Africa and in parts of central Chile. The climate is characterized by hot, dry summers and cool, wet winters.

A steppe is a dry grassland with an annual temperature range in the summer of up to 40 °C (104 °F) and during the winter down to "40 °C ("40 °F).

A subarctic climate has little precipitation, and monthly temperatures which are above 10 °C (50 °F) for one to three months of the year, with permafrost in large parts of the area due to the cold winters. Winters within subarctic climates usually include up to six months of temperatures averaging below 0 °C (32 °F).

A polar ice cap, or polar ice sheet, is a high-latitude region of a planet or moon that is covered in ice. Ice caps form because high-latitude regions receive less energy as solar radiation from the sun than equatorial regions, resulting in lower surface temperatures.

A desert is a landscape form or region that receives very little precipitation. Deserts usually have a large diurnal and seasonal temperature range, with high or low, depending on location daytime temperatures (in summer up to 45 °C or 113 °F), and low nighttime temperatures (in winter down to 0 °C or 32 °F) due to extremely low humidity. Many deserts are formed by rain shadows, as mountains block the path of moisture and precipitation to the desert.

Thornthwaite

Devised by the American climatologist and geographer C. W. Thornthwaite, this climate classification method monitors the soil water budget using evapotranspiration. It monitors the portion of total precipitation used to nourish vegetation over a certain area. It uses indices such as a humidity index and an aridity index to determine an area's moisture regime based upon its average temperature, average rainfall, and average vegetation type. The lower the value of the index in any given area, the drier the area is.

The moisture classification includes climatic classes with descriptors such as hyperhumid, humid, subhumid, subarid, semi-arid (values of "20 to "40), and arid (values below "40). Humid regions experience more precipitation than evaporation each year, while arid regions experience greater evaporation than precipitation on an annual basis. A total of 33 percent of the Earth's landmass is considered either arid of semi-arid, including southwest North America, southwest South America, most of northern and a small part of southern Africa, southwest and portions of eastern Asia, as well as much of Australia. Studies suggest that precipitation effectiveness (PE) within the Thornthwaite moisture index is overestimated in the summer and underestimated in the winter. This index can be effectively used to determine the number of herbivore and mammal species numbers within a given area. The index is also used in studies of climate change.

Thermal classifications within the Thornthwaite scheme include microthermal, mesothermal, and megathermal regimes. A microthermal climate is one of low annual mean temperatures, generally between 0 °C (32 °F) and 14 °C (57 °F) which experiences short summers and has a potential evaporation between 14 centimetres (5.5 in) and 43 centimetres (17 in). A mesothermal climate lacks persistent heat or persistent cold, with potential evaporation between 57 centimetres (22 in) and 114 centimetres (45 in). A megathermal climate is one with persistent high temperatures and abundant rainfall, with potential annual evaporation in excess of 114 centimetres (45 in).

Record

Modern :

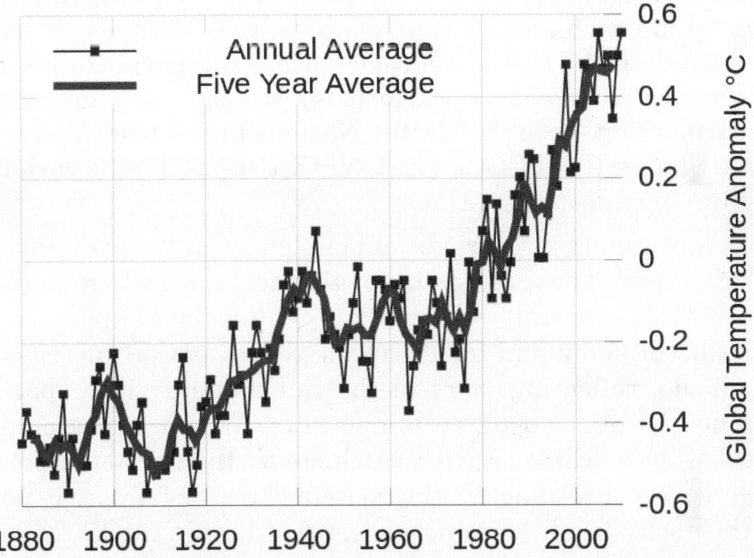

Figure: *Instrumental temperature record of the last 150 years*

Details of the modern climate record are known through the taking of measurements from such weather instruments as thermometers, barometers, and anemometers during the past few centuries. The instruments used to study weather over the modern time scale, their known error, their immediate environment, and their exposure have changed over the years, which must be considered when studying the climate of centuries past.

Paleoclimatology

Paleoclimatology is the study of past climate over a great period of the Earth's history. It uses evidence from ice sheets, tree rings, sediments, coral, and rocks to determine the past state of the climate.

It demonstrates periods of stability and periods of change and can indicate whether changes follow patterns such as regular cycles.

Climate Change

Climate change is the variation in global or regional climates over time. It reflects changes in the variability or average state of the atmosphere over time scales ranging from decades to millions of years. These changes can be caused by processes internal to the Earth, external forces (e.g. variations in sunlight intensity) or, more recently, human activities.

In recent usage, especially in the context of environmental policy, the term "climate change" often refers only to changes in modern climate, including the rise in average surface temperature known as global warming. In some cases, the term is also used with a presumption of human causation, as in the United Nations Framework Convention on Climate Change (UNFCCC). The UNFCCC uses "climate variability" for non-human caused variations.

Earth has undergone periodic climate shifts in the past, including four major ice ages. These consisting of glacial periods where conditions are colder than normal, separated by interglacial periods. The accumulation of snow and ice during a glacial period increases the surface albedo, reflecting more of the Sun's energy into space and maintaining a lower atmospheric temperature. Increases in greenhouse gases, such as by volcanic activity, can increase the global temperature and produce an interglacial. Suggested causes of ice age periods include the positions of the continents, variations in the Earth's orbit, changes in the solar output, and volcanism.

Climate Models

Climate models use quantitative methods to simulate the interactions of the atmosphere, oceans, land surface and ice. They are used for a variety of purposes; from the study of the dynamics of the weather and climate system, to projections of future climate. All climate models balance, or very nearly balance, incoming energy as short wave (including visible) electromagnetic radiation to the earth with outgoing energy as long wave (infrared) electromagnetic radiation from the earth. Any imbalance results in a change in the average temperature of the earth.

The most talked-about applications of these models in recent years have been their use to infer the consequences of increasing greenhouse gases in the atmosphere, primarily carbon dioxide. These models predict an upward trend in the global mean surface temperature, with the most rapid increase in temperature being projected for the higher latitudes of the Northern Hemisphere.

Models can range from relatively simple to quite complex:

- Simple radiant heat transfer model that treats the earth as a single point and averages outgoing energy
- this can be expanded vertically (radiative-convective models), or horizontally
- finally, (coupled) atmosphere–ocean–sea ice global climate models discretise and solve the full equations for mass and energy transfer and radiant exchange.

Climate forecasting is a way by some scientists are using to predict climate change. In 1997 the prediction division of the International Research Institute for Climate and Society at Columbia University began generating seasonal climate forecasts on a real-time basis. To produce these forecasts an extensive suite of forecasting tools was developed, including a multimodel ensemble approach that required thorough validation of each model's accuracy level in simulating interannual climate variability

Alliance for Climate Protection

The Alliance for Climate Protection was founded in 2006 by Nobel laureate and former United States Vice President Al Gore.

The Alliance is a nonprofit, nonpartisan organization committed to educating the global community about the urgency of implementing comprehensive solutions to the climate crisis. The organization oversees numerous campaigns, including Repower America, the Climate Project, Repower at Home, the WE Campaign, the Reality Coalition, and works with its affiliated organization, the Climate Protection Action Fund.

The Alliance emphasizes the need to draw together a diverse range of individuals and organizations, including representatives of labour, faith, business, schools, women, agriculture, sportsmen and many other communities. According to the website, the Alliance "brings together faces and voices in a way that diversifies and strengthens the global network of concerned individuals who want to take action now on climate issues."

Leadership

Maggie L. Fox, President and CEO, is a veteran of numerous national issues, political and environmental campaigns and has spent 30 years working to mobilize Americans to take action. Maggie is the past National President of America Votes, a progressive coalition of over 40 organizations spearheading the largest voter mobilization and education effort in the nation. She served as the Deputy Executive Director of the Sierra Club, the nation's oldest and largest grassroots environmental advocacy organization. Most recently, Maggie consulted with a number of organizations on their energy and climate campaigns including the Energy Future Coalition, Western Resource Advocates, and the Ocean Conservancy.

Maggie began her career as a teacher and community organizer on the Navajo and Hopi Reservations of Arizona and New Mexico and worked for the Colorado, North Carolina and Northwest Outward Bound Schools for over a decade. She earned her B.A. from the University of North Carolina, a Masters in Education from the University of Colorado, and a J.D. with an emphasis in Environmental Law and Native America Natural Resources Law from Northwestern School of Law.

Funding

Much of the Alliance's funding comes from Al Gore's work. Paramount Classics, distributor of Gore's documentary *An Inconvenient Truth* announced in August 2006 that it would donate 5% of all box office receipts to the Alliance. Al Gore is also donating all his proceeds from the film to the Alliance. After receiving the 2007 Nobel Peace Prize, Al Gore pledged to donate 100% of his share of the prize (approximately $1.5 million) to the Alliance. Gore's recently published book, *Our Choice*, have had all proceeds donated to the Alliance.

Some of the profits from the Live Earth concerts, and profits from the associated book - *The Live Earth Global Warming Survival Handbook*, will also be used to fund the alliance, which is also seeking contributions from other donors.

Projects

Repower America: Repower America is a campaign launched in 2008 by Al Gore, based on a speech he gave in July 2008. The campaigns calls for a plan to "repower" the United States and revitalize national energy infrastructure, through investment in energy efficiency, clean renewable energy sources, a unified smart energy national grid and clean car technology.

A major effort of the Repower America campaign is the Repower Wall, an interactive online forum housing personal videos and written testimonials calling for action to solve the climate crisis. The thousands of contributors to the Wall have been featured in Repower television advertising.

The Repower America brand is shared by the Alliance and by affiliate the Climate Protection Action Fund.

The Climate Project

The Climate Project (TCP) educates and engages people at a peer-to-peer level worldwide about the urgency and solvability of the climate crisis. The Presenters—volunteers trained to give a version of Vice President Al Gore's slideshow featured in *An Inconvenient Truth* — are active in 55 countries and have given a combined total of over 70,000 presentations. The Alliance for Climate Protection and The Climate Project unified their programmes and activities in March 2010.

Repower at Home

In 2010, the Alliance for Climate Protection launched Repower at Home, a national programme dedicated to reducing residential energy waste.

Repower at Home provides users with information about saving energy in their own homes, and tools for inspiring others to follow their lead. Repower at Home has challenged Americans to reduce their energy use by the equivalent of seven million pounds of coal by October 10, 2010.

WE Campaign

In March 2008 the Alliance launched a three-year, $300 million campaign aimed at mobilizing Americans to push for aggressive reductions in greenhouse gas emissions. This campaign is one of the most ambitious and costly public advocacy campaigns in U.S. history. In August 2008, the We Campaign released a television advertisement demanding that American leaders give the American people "truly clean energy" within ten years (Alliance for Climate Protection's Repower America project)

Reality Coalition

The Alliance for Climate Protection, League of Conservation Voters, National Wildlife Federation, Natural Resources Defence Council and Sierra Club launched the Reality Coalition, a national grassroots and an aggressive multimillion-dollar advertising campaign to change the public perception of clean coal. In early 2009 the Coen Brothers directed a TV commercial for the group titled *This Is Reality*.

Affiliates

The Climate Protection Action Fund is a non-profit, non-partisan affiliated organization that directly influences the passage of key local, state and federal climate and energy legislation and policies that are critical to solving our most urgent climate problems. According to the website, the Action Fund uses using targeted legislative campaigns, issue advocacy and grassroots organizing to "represent the powerful combined voices of the millions of Americans who understand that global warming is real and that the time to act is now."

Greenhouse Gas Emissions From Agricultural Fields Making Agriculture Truly Environment Friendly

A greenhouse gas (sometimes abbreviated GHG) is a gas in an atmosphere that absorbs and emits radiation within the thermal infrared range. This process is the fundamental cause of the greenhouse effect. The primary greenhouse gases in the Earth's atmosphere are water vapour, carbon dioxide, methane, nitrous oxide, and ozone. In the Solar System, the atmospheres of Venus, Mars, and Titan also contain gases that cause greenhouse effects. Greenhouse gases greatly affect the temperature of the Earth; without them, Earth's surface would average about 33°C colder than the present average of 14 °C (57 °F).

Since the beginning of the Industrial Revolution, the burning of fossil fuels has contributed to a 40% increase in the concentration of carbon dioxide in the atmosphere from 280 ppm to 397 ppm, despite the uptake of a large portion of the emissions by various natural "sinks" involved in the carbon cycle. Anthropogenic carbon dioxide (CO_2) emissions (i.e., emissions produced by human activities) come from combustion of carbon based fuels, principally wood, coal, oil, and natural gas.

Gases in Earth's Atmosphere

Greenhouse gases :

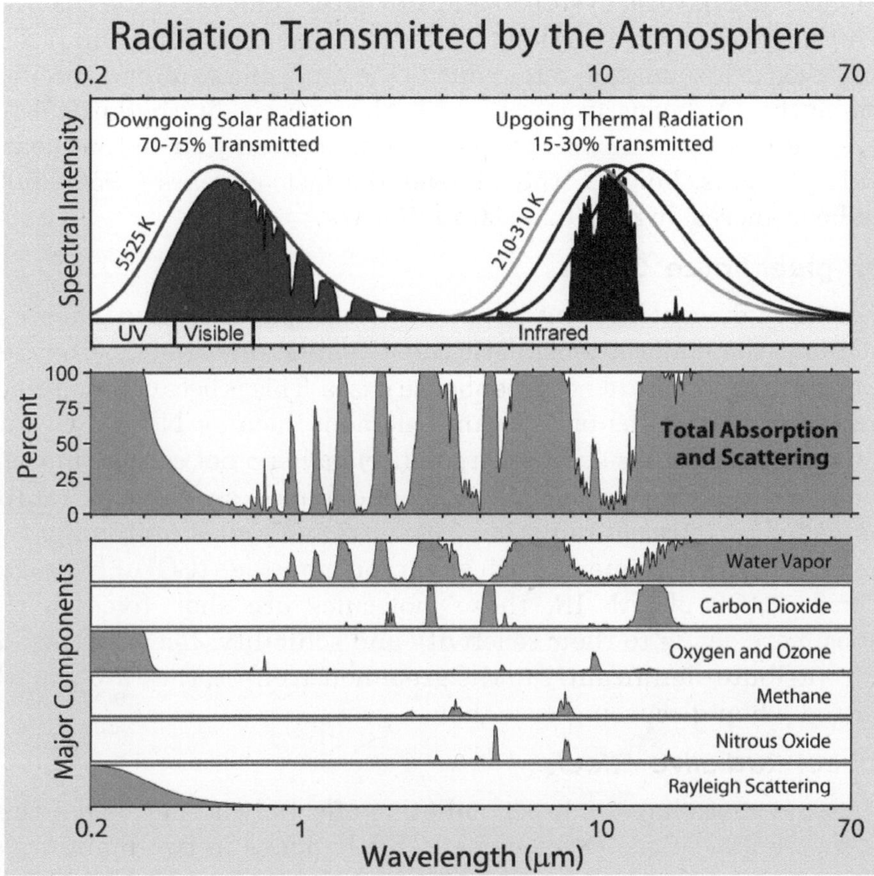

Figure: *Atmospheric absorption and scattering at different wavelengths of electromagnetic waves. The largest absorption band of carbon dioxide is in the infrared.*

Greenhouse gases are those that can absorb and emit infrared radiation, but not radiation in or near the visible spectrum. In order, the most abundant greenhouse gases in Earth's atmosphere are:

- water vapour (H_2O)
- carbon dioxide (CO_2)
- methane (CH_4)
- nitrous oxide (N_2O)
- ozone (O_3)

Atmospheric concentrations of greenhouse gases are determined by the balance between sources (emissions of the gas from human activities and natural systems) and sinks (the removal of the gas from the atmosphere by conversion to a different chemical compound). The proportion of an emission remaining in the atmosphere after a specified time is the "Airborne fraction" (AF). More precisely, the annual AF is the ratio of the atmospheric increase in a given year to that year's total emissions. For CO_2 the AF over the last 50 years (1956–2006) has been increasing at 0.25 ± 0.21%/year.

Non-greenhouse Gases

Although contributing to many other physical and chemical reactions, the major atmospheric constituents, nitrogen (N_2), oxygen (O_2), and argon (Ar), are not greenhouse gases. This is because molecules containing two atoms of the same element such as N_2 and O_2 and monatomic molecules such as Argon (Ar) have no net change in their dipole moment when they vibrate and hence are almost totally unaffected by infrared radiation. Although molecules containing two atoms of different elements such as carbon monoxide (CO) or hydrogen chloride (HCl) absorb IR, these molecules are short-lived in the atmosphere owing to their reactivity and solubility. Because they do not contribute significantly to the greenhouse effect, they are usually omitted when discussing greenhouse gases.

Indirect Radiative Effects

Some gases have indirect radiative effects (whether or not they are a greenhouse gas themselves). This happens in two main ways. One way is that when they break down in the atmosphere they produce another greenhouse gas. For example methane and carbon monoxide (CO) are oxidised to give carbon dioxide (and methane oxidation also produces water vapour; that will be considered below). Oxidation of CO to CO_2 directly produces an unambiguous increase in radiative forcing although the reason is subtle. The peak of the thermal IR emission from the Earth's surface is very close to a strong vibrational absorption band of CO_2 (667 cm)$^{-1}$. On the other hand, the single CO vibrational band only absorbs IR at much higher frequencies (2145 cm^{-1}), where the ~300K thermal emission of the surface is at least a factor of ten lower. On the other hand, oxidation of methane to CO_2 which requires reactions with the OH radical, produces an instantaneous reduction, since CO_2 is a weaker greenhouse gas than methane; but it has a longer lifetime. As described below this is not the whole story, since the oxidations of CO and CH_4 are intertwined by both consuming OH radicals. In any case, the calculation of the total radiative effect needs to include both the direct and indirect forcing.

A second type of indirect effect happens when chemical reactions in the atmosphere involving these gases change the concentrations of greenhouse gases. For example, the destruction of non methane volatile organic compounds (NMVOC) in the atmosphere can produce ozone. The size of the indirect effect can depend strongly on where and when the gas is emitted.

Methane has a number of indirect effects in addition to forming CO_2. Firstly, the main chemical which destroys methane in the atmosphere is the hydroxyl radical (OH). Methane reacts with OH and so more methane means that the concentration of OH goes down. Effectively, methane increases its own atmospheric lifetime and therefore its overall radiative effect. The second effect is that the oxidation of methane can produce ozone. Thirdly, as well as making CO_2 the oxidation of methane produces water; this is a major source of water vapour in the stratosphere which is otherwise very dry. CO and NMVOC also produce CO_2 when they are oxidised. They remove OH from the atmosphere and this leads to higher concentrations of methane. The surprising effect of this is that the global warming potential of CO is three times that of CO_2. The same process that converts NMVOC to carbon dioxide can also lead to the formation of tropospheric ozone. Halocarbons have an indirect effect because they destroy stratospheric ozone. Finally hydrogen can lead to ozone production and CH_4 increases as well as producing water vapour in the stratosphere.

Contribution of Clouds to Earth's Greenhouse Effect

The major non-gas contributor to the Earth's greenhouse effect, clouds, also absorb and emit infrared radiation and thus have an effect on radiative properties of the greenhouse gases. Clouds are water droplets or ice crystals suspended in the atmosphere.

Impacts on the Overall Greenhouse Effect

Schmidt *et al.* (2010) analysed how individual components of the atmosphere contribute to the total greenhouse effect. They estimated that water vapour accounts for about 50% of the Earth's greenhouse effect, with clouds contributing 25%, carbon dioxide 20%, and the minor greenhouse gases and aerosols accounting for the remaining 5%. In the study, the reference model atmosphere is for 1980 conditions. Image credit: NASA.

The contribution of each gas to the greenhouse effect is affected by the characteristics of that gas, its abundance, and any indirect effects it may cause. For example, the direct radiative effect of a mass of methane is about 72 times stronger than the same mass of carbon dioxide over a 20 year time frame but it is present in much smaller concentrations so that its total direct radiative effect is smaller, in part due to its shorter atmospheric lifetime.

On the other hand, in addition to its direct radiative impact, methane has a large, indirect radiative effect because it contributes to ozone formation. Shindell *et al.* (2005) argue that the contribution to climate change from methane is at least double previous estimates as a result of this effect.

When ranked by their direct contribution to the greenhouse effect, the most important are:

Compound	*Formula*	*Contribution (%)*
Water vapor and clouds	H_2O	36 – 72%
Carbon dioxide	CO_2	9 – 26%
Methane	CH_4	4 – 9%
Ozone	O_3	3 – 7%

In addition to the main greenhouse gases listed above, other greenhouse gases include sulfur hexafluoride, hydrofluorocarbons and perfluorocarbons. Some greenhouse gases are not often listed. For example, nitrogen trifluoride has a high global warming potential (GWP) but is only present in very small quantities.

Proportion of Direct Effects at a Given Moment

It is not possible to state that a certain gas causes an exact percentage of the greenhouse effect. This is because some of the gases absorb and emit radiation at the same frequencies as others, so that the total greenhouse effect is not simply the sum of the influence of each gas. The higher ends of the ranges quoted are for each gas alone; the lower ends account for overlaps with the other gases. In addition, some gases such as methane are known to have large indirect effects that are still being quantified.

Radiative Forcing

The Earth absorbs some of the radiant energy received from the sun, reflects some of it as light and reflects or re-radiates the rest back to space as heat. The Earth's surface temperature depends on this balance between incoming and outgoing energy. If this energy balance is shifted, the Earth's surface could become warmer or cooler, leading to a variety of changes in global climate.

A number of natural and man-made mechanisms can affect the global energy balance and force changes in the Earth's climate. Greenhouse gases are one such mechanism.

Greenhouse gases in the atmosphere absorb and re-emit some of the outgoing energy radiated from the Earth's surface, causing that heat to be retained in the lower atmosphere.

As explained above, some greenhouse gases remain in the atmosphere for decades or even centuries, and therefore can affect the Earth's energy balance over a long time period.

Factors that influence Earth's energy balance can be quantified in terms of "radiative climate forcing." Positive radiative forcing indicates warming (for example, by increasing incoming energy or decreasing the amount of energy that escapes to space), while negative forcing is associated with cooling.

Global Warming Potential

The global warming potential (GWP) depends on both the efficiency of the molecule as a greenhouse gas and its atmospheric lifetime. GWP is measured relative to the same mass of CO_2 and evaluated for a specific timescale. Thus, if a gas has a high (positive) radiative forcing but also a short lifetime, it will have a large GWP on a 20 year scale but a small one on a 100 year scale. Conversely, if a molecule has a longer atmospheric lifetime than CO_2 its GWP will increase with the timescale considered. Carbon dioxide is defined to have a GWP of 1 over all time periods.

Methane has an atmospheric lifetime of 12 ± 3 years and a GWP of 72 over 20 years, 25 over 100 years and 7.6 over 500 years. The decrease in GWP at longer times is because methane is degraded to water and CO_2 through chemical reactions in the atmosphere.

Examples of the atmospheric lifetime and GWP relative to CO_2 for several greenhouse gases are given in the following table:

Atmospheric lifetime and GWP relative to CO_2 at different time horizon for various greenhouse gases.					
Gas name	***Chemical formula***	***Lifetime (years)***	***Global warming potential (GWP) for given time horizon***		
			20-yr	***100-yr***	***500-yr***
Carbon dioxide	CO_2	See above	1	1	1
Methane	CH_4	12	72	25	7.6
Nitrous oxide	N_2O	114	289	298	153
CFC-12	CCl_2F_2	100	11 000	10 900	5 200
HCFC-22	$CHClF_2$	12	5 160	1 810	549
Tetrafluoromethane	CF_4	50 000	5 210	7 390	11 200
Hexafluoroethane	C_2F_6	10 000	8 630	12 200	18 200
Sulfur hexafluoride	SF_6	3 200	16 300	22 800	32 600
Nitrogen trifluoride	NF_3	740	12 300	17 200	20 700

The use of CFC-12 (except some essential uses) has been phased out due to its ozone depleting properties. The phasing-out of less active HCFC-compounds will be completed in 2030.

Natural and Anthropogenic Sources

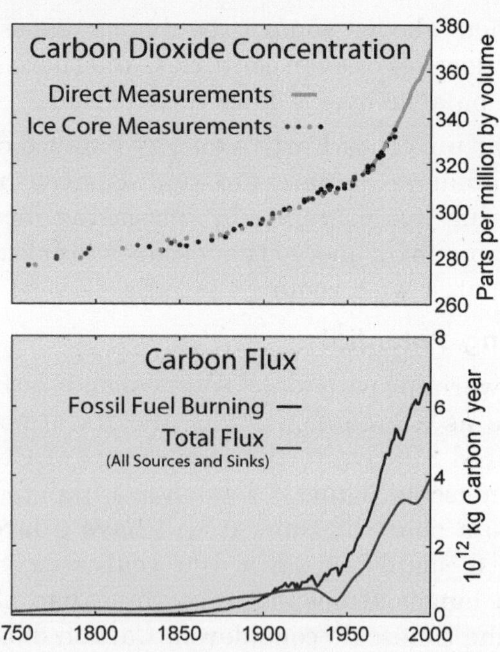

Figure: *Top: Increasing atmospheric carbon dioxide levels as measured in the atmosphere and reflected in ice cores. Bottom: The amount of net carbon increase in the atmosphere, compared to carbon emissions from burning fossil fuel.*

Aside from purely human-produced synthetic halocarbons, most greenhouse gases have both natural and human-caused sources. During the pre-industrial Holocene, concentrations of existing gases were roughly constant. In the industrial era, human activities have added greenhouse gases to the atmosphere, mainly through the burning of fossil fuels and clearing of forests.

The 2007 Fourth Assessment Report compiled by the IPCC (AR4) noted that "changes in atmospheric concentrations of greenhouse gases and aerosols, land cover and solar radiation alter the energy balance of the climate system", and concluded that "increases in anthropogenic greenhouse gas concentrations is very likely to have caused most of the increases in global average temperatures since the mid-20th century". In AR4, "most of" is defined as more than 50%.

Abbreviations used in the two tables below: ppm = parts-per-million; ppb = parts-per-billion; ppt = parts-per-trillion; W/m² = watts per square metre

***Table:** Current greenhouse gas concentrations*

Gas	*Pre-1750 tropospheric concentration*	*Recent tropospheric concentration*	*Absolute increase since 1750*	*Percentage increase since 1750*	*Increased radiative forcing (W/m^2)*
Carbon dioxide (CO_2)	280 ppm	392.6 ppm	112.6 ppm	40.2%	1.85
Methane (CH_4)	700 ppb	1874 ppb / 1758 ppb	1174 ppb / 1058 ppb	167.7% / 151.1%	0.51
Nitrous oxide (N_2O)	270 ppb	324 ppb / 323 ppb	54 ppb / 53 ppb	20.0% / 19.6%	0.18
Tropospheric ozone (O_3)	25 ppb	34 ppb	9 ppb	36%	0.35

Relevant to radiative forcing and/or ozone depletion; all of the following have no natural sources and hence zero amounts pre-industrial.

Ice cores provide evidence for greenhouse gas concentration variations over the past 800,000 years. Both CO_2 and CH_4 vary between glacial and interglacial phases, and concentrations of these gases correlate strongly with temperature. Direct data does not exist for periods earlier than those represented in the ice core record, a record that indicates CO_2 mole fractions stayed within a range of 180 ppm to 280 ppm throughout the last 800,000 years, until the increase of the last 250 years. However, various proxies and modelling suggests larger variations in past epochs; 500 million years ago CO_2 levels were likely 10 times higher than now. Indeed higher CO_2 concentrations are thought to have prevailed throughout most of the Phanerozoic eon, with concentrations four to six times current concentrations during the Mesozoic era, and ten to fifteen times current concentrations during the early Palaeozoic era until the middle of the Devonian period, about 400 Ma. The spread of land plants is thought to have reduced CO_2 concentrations during the late Devonian, and plant activities as both sources and sinks of CO_2 have since been important in providing stabilising feedbacks.

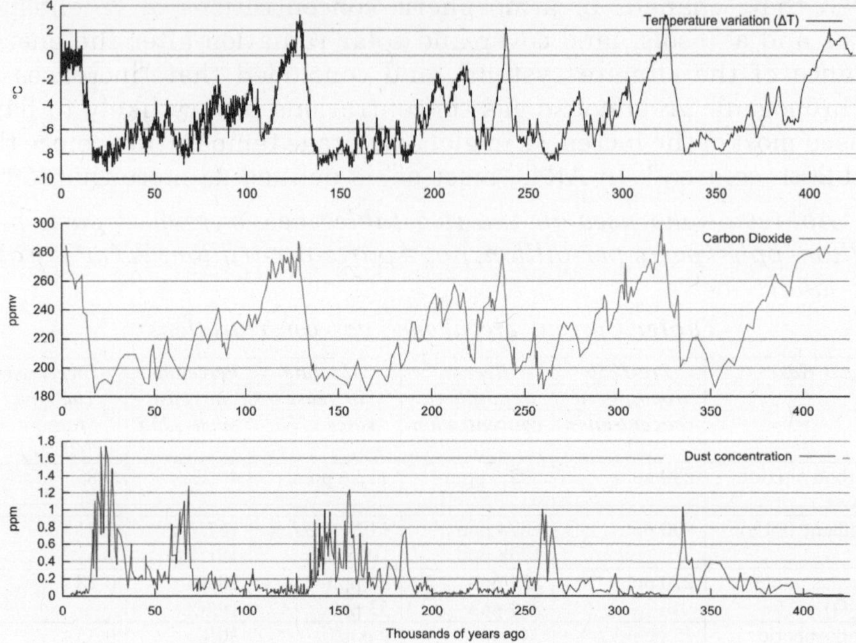

***Figure:** 400,000 years of ice core data.*

Earlier still, a 200-million year period of intermittent, widespread glaciation extending close to the equator (Snowball Earth) appears to have been ended suddenly, about 550 Ma, by a colossal volcanic outgassing that raised the CO_2 concentration of the atmosphere abruptly to 12%, about 350 times modern levels, causing extreme greenhouse conditions and carbonate deposition as limestone at the rate of about 1 mm per day. This episode marked the close of the Precambrian eon, and was succeeded by the generally warmer conditions of the Phanerozoic, during which multicellular animal and plant life evolved. No volcanic carbon dioxide emission of comparable scale has occurred since. In the modern era, emissions to the atmosphere from volcanoes are only about 1% of emissions from human sources.

Ice Cores

Measurements from Antarctic ice cores show that before industrial emissions started atmospheric CO_2 mole fractions were about 280 parts per million (ppm), and stayed between 260 and 280 during the preceding ten thousand years. Carbon dioxide mole fractions in the atmosphere have gone up by approximately 35 percent since the 1900s, rising from 280 parts per million by volume to 387 parts per million in 2009. One study using evidence from stomata of fossilized leaves suggests greater variability, with carbon dioxide mole fractions above 300 ppm during the period seven to ten thousand years ago, though others have argued that these findings more likely reflect calibration or contamination problems rather than actual CO_2 variability. Because of the way air is trapped in ice (pores in the ice close off slowly to form bubbles deep within the firn) and the time period represented in each ice sample analyzed, these figures represent averages of atmospheric concentrations of up to a few centuries rather than annual or decadal levels.

Changes Since the Industrial Revolution

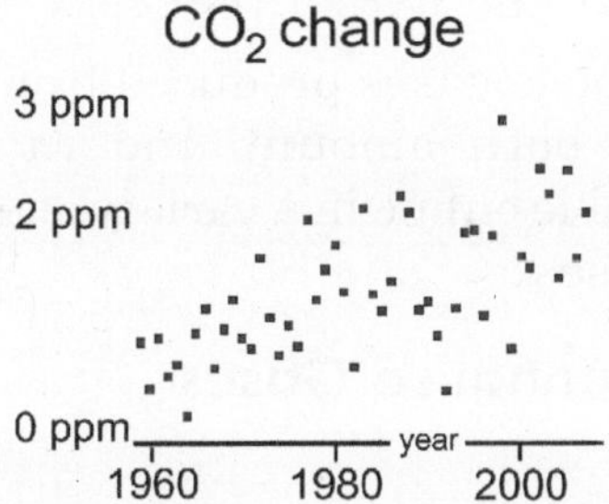

Figure: *Recent year-to-year increase of atmospheric* CO_2.

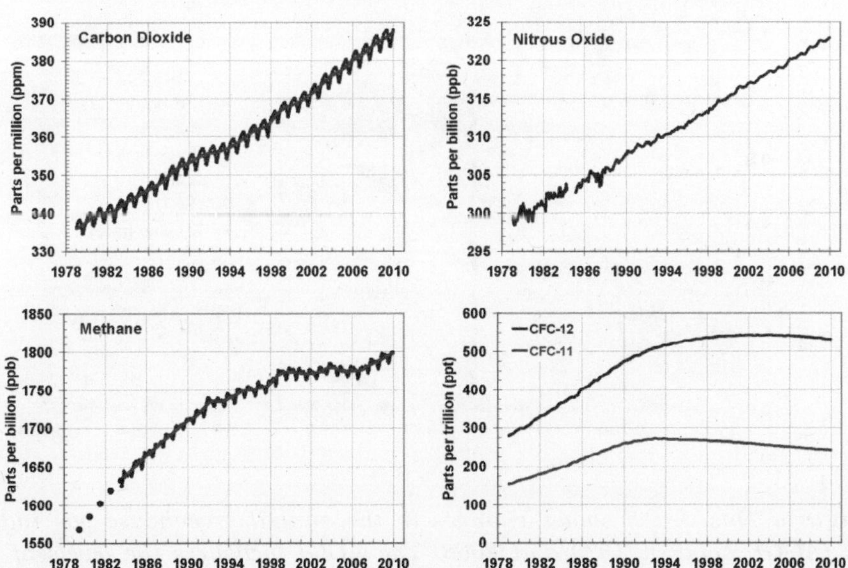

Figure: *Major greenhouse gas trends.*

Since the beginning of the Industrial Revolution, the concentrations of most of the greenhouse gases have increased. For example, the mole fraction of carbon dioxide has increased from 280 ppm by about 36% to 380 ppm, or 100 ppm over modern pre-industrial levels. The first 50 ppm increase took place in about 200 years, from the start of the Industrial Revolution to around 1973; however the next 50 ppm increase took place in about 33 years, from 1973 to 2006.

Recent data also shows that the concentration is increasing at a higher rate. In the 1960s, the average annual increase was only 37% of what it was in 2000 through 2007. Today, the stock of carbon in the atmosphere increases by more than 3 million tonnes per annum (0.04%) compared with the existing stock. This increase is the result of human activities by burning fossil fuels, deforestation and forest degradation in tropical and boreal regions.

The other greenhouse gases produced from human activity show similar increases in both amount and rate of increase. Many observations are available online in a variety of Atmospheric Chemistry Observational Databases.

Anthropogenic Greenhouse Gases

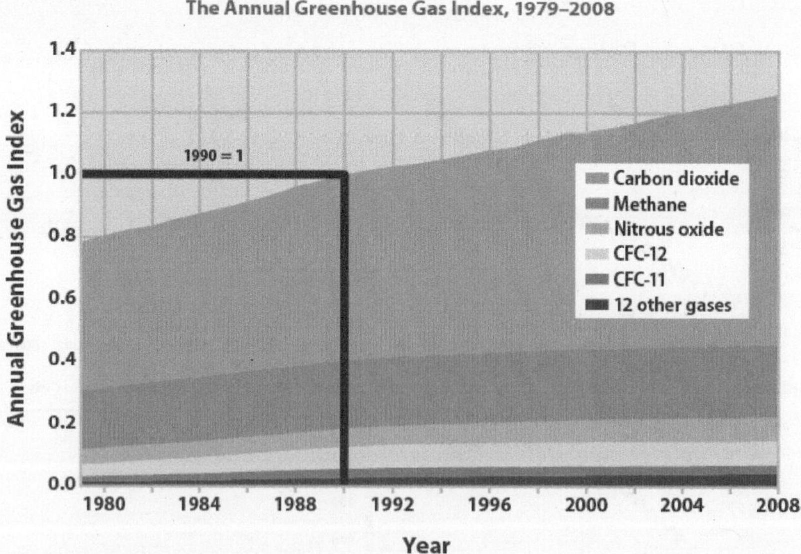

Figure: *This graph shows changes in the annual greenhouse gas index (AGGI) between 1979 and 2008. The AGGI measures the levels of greenhouse gases in the atmosphere based on their ability to cause changes in the Earth's climate.*

This bar graph shows global greenhouse gas emissions by sector from 1990 to 2005, measured in carbon dioxide equivalents.

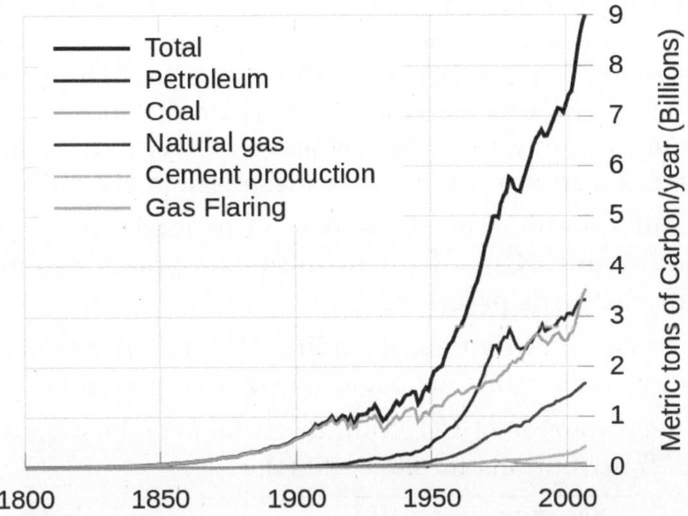

Figure: *Modern global anthropogenic carbon emissions.*

Since about 1750 human activity has increased the concentration of carbon dioxide and other greenhouse gases. Measured atmospheric concentrations of carbon dioxide are currently 100 ppm higher than pre-industrial levels. Natural sources of carbon dioxide are more than 20 times greater than sources due to human activity, but over periods longer than a few years natural sources are closely balanced by natural sinks, mainly photosynthesis of carbon compounds by plants and marine plankton. As a result of this balance, the atmospheric mole fraction of carbon dioxide remained between 260 and 280 parts per million for the 10,000 years between the end of the last glacial maximum and the start of the industrial era.

It is likely that anthropogenic (i.e., human-induced) warming, such as that due to elevated greenhouse gas levels, has had a discernible influence on many physical and biological systems. Future warming is projected to have a range of impacts, including sea level rise, increased frequencies and severities of some extreme weather events, loss of biodiversity, and regional changes in agricultural productivity.

The main sources of greenhouse gases due to human activity are:

- burning of fossil fuels and deforestation leading to higher carbon dioxide concentrations in the air. Land use change (mainly deforestation in the tropics) account for up to one third of total anthropogenic CO_2 emissions.

- livestock enteric fermentation and manure management, paddy rice farming, land use and wetland changes, pipeline losses, and covered vented landfill emissions leading to higher methane atmospheric concentrations. Many of the newer style fully vented septic systems that enhance and target the fermentation process also are sources of atmospheric methane.
- use of chlorofluorocarbons (CFCs) in refrigeration systems, and use of CFCs and halons in fire suppression systems and manufacturing processes.
- agricultural activities, including the use of fertilizers, that lead to higher nitrous oxide (N_2O) concentrations.

The seven sources of CO_2 from fossil fuel combustion are (with percentage contributions for 2000–2004):

Seven main fossil fuel combustion sources	***Contribution (%)***
Liquid fuels (e.g., gasoline, fuel oil)	36%
Solid fuels (e.g., coal)	35%
Gaseous fuels (e.g., natural gas)	20%
Cement production	3 %
Flaring gas industrially and at wells	< 1%
Non-fuel hydrocarbons	< 1%
"International bunker fuels" of transport not included in national inventories	4 %

Carbon dioxide, methane, nitrous oxide (N_2O) and three groups of fluorinated gases (sulfur hexafluoride (SF_6), hydrofluorocarbons (HFCs), and perfluorocarbons (PFCs)) are the major anthropogenic greenhouse gases, and are regulated under the Kyoto Protocol international treaty, which came into force in 2005. Emissions limitations specified in the Kyoto Protocol expire in 2012. The Cancún agreement, agreed in 2010, includes voluntary pledges made by 76 countries to control emissions. At the time of the agreement, these 76 countries were collectively responsible for 85% of annual global emissions.

Although CFCs are greenhouse gases, they are regulated by the Montreal Protocol, which was motivated by CFCs' contribution to ozone depletion rather than by their contribution to global warming.

Note that ozone depletion has only a minor role in greenhouse warming though the two processes often are confused in the media.

Role of Water Vapour

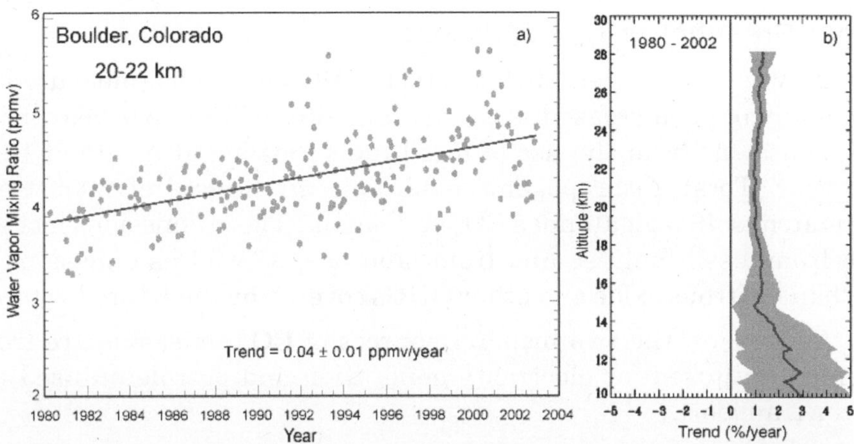

***Figure:** Increasing water vapour in the stratosphere at Boulder, Colorado.*

Water vapour accounts for the largest percentage of the greenhouse effect, between 36% and 66% for clear sky conditions and between 66% and 85% when including clouds. Water vapour concentrations fluctuate regionally, but human activity does not significantly affect water vapour concentrations except at local scales, such as near irrigated fields. The atmospheric concentration of vapour is highly variable and depends largely on temperature, from less than 0.01% in extremely cold regions up to 3% by mass at in saturated air at about 32 °C.

The average residence time of a water molecule in the atmosphere is only about nine days, compared to years or centuries for other greenhouse gases such as CH_4 and CO_2. Thus, water vapour responds to and amplifies effects of the other greenhouse gases. The Clausius-Clapeyron relation establishes that more water vapour will be present per unit volume at elevated temperatures. This and other basic principles indicate that warming associated with increased concentrations of the other greenhouse gases also will increase the concentration of water vapour (assuming that the relative humidity remains approximately constant; modelling and observational studies find that this is indeed so). Because water vapour is a greenhouse gas, this results in further warming and so is a "positive feedback" that amplifies the original warming. Eventually other earth processes offset these positive feedbacks, stabilizing the global temperature at a new equilibrium and preventing the loss of Earth's water through a Venus-like runaway greenhouse effect.

Direct Greenhouse Gas Emissions

Between the period 1970 to 2004, GHG emissions (measured in CO_2-equivalent) increased at an average rate of 1.6% per year, with CO_2 emissions from the use of fossil fuels growing at a rate of 1.9% per year. Total anthropogenic emissions at the end of 2009 were estimated at 49.5 gigatonnes CO_2-equivalent. These emissions include CO_2 from fossil fuel use and from land use, as well as emissions of methane, nitrous oxide and other GHGs covered by the Kyoto Protocol.

At present, the two primary sources of CO_2 emissions are from burning coal used for electricity generation and petroleum used for motor transport.

Regional and National Attribution of Emissions

There are several different ways of measuring GHG emissions, for example, World Bank (2010) for tables of national emissions data. Some variables that have been reported include:

- Definition of measurement boundaries: Emissions can be attributed geographically, to the area where they were emitted (the territory principle) or by the activity principle to the territory produced the emissions. These two principles result in different totals when measuring, for example, electricity importation from one country to another, or emissions at an international airport.
- Time horizon of different GHGs: Contribution of a given GHG is reported as a CO_2 equivalent. The calculation to determine this takes into account how long that gas remains in the atmosphere. This is not always known accurately and calculations must be regularly updated to reflect new information.
- What sectors are included in the calculation (e.g., energy industries, industrial processes, agriculture etc.): There is often a conflict between transparency and availability of data.
- The measurement protocol itself: This may be via direct measurement or estimation. The four main methods are the

emission factor-based method, mass balance method, predictive emissions monitoring systems, and continuous emissions monitoring systems. These methods differ in accuracy, cost, and usability.

These different measures are sometimes used by different countries to assert various policy/ethical positions on climate change (Banuri *et al.*, 1996, p. 94). This use of different measures leads to a lack of comparability, which is problematic when monitoring progress towards targets. There are arguments for the adoption of a common measurement tool, or at least the development of communication between different tools.

Emissions may be measured over long time periods. This measurement type is called historical or cumulative emissions. Cumulative emissions give some indication of who is responsible for the build-up in the atmospheric concentration of GHGs (IEA, 2007, p. 199).

The national accounts balance would be positively related to carbon emissions. The national accounts balance shows the difference between exports and imports. For many richer nations, such as the United States, the accounts balance is negative because more goods are imported than they are exported. This is mostly due to the fact that it is cheaper to produce goods outside of developed countries, leading the economies of developed countries to become increasingly dependent on services and not goods. We believed that a positive accounts balance would means that more production was occurring in a country, so more factories working would increase carbon emission levels.(Holtz-Eakin, 1995, pp.;85;101).

Emissions may also be measured across shorter time periods. Emissions changes may, for example, be measured against a base year of 1990. 1990 was used in the United Nations Framework Convention on Climate Change (UNFCCC) as the base year for emissions, and is also used in the Kyoto Protocol (some gases are also measured from the year 1995). A country's emissions may also be reported as a proportion of global emissions for a particular year.

Another measurement is of per capita emissions. This divides a country's total annual emissions by its mid-year population. Per capita emissions may be based on historical or annual emissions (Banuri *et al.*, 1996, pp. 106–107).

Greenhouse Gas Intensity and Land-use Change

Land-use change, e.g., the clearing of forests for agricultural use, can affect the concentration of GHGs in the atmosphere by altering how much carbon flows out of the atmosphere into carbon sinks. Accounting for land-use change can be understood as an attempt to measure "net" emissions, i.e., gross emissions from all GHG sources minus the removal of emissions from the atmosphere by carbon sinks (Banuri *et al.*, 1996, pp. 92–93).

There are substantial uncertainties in the measurement of net carbon emissions. Additionally, there is controversy over how carbon sinks should be allocated between different regions and over time (Banuri *et al.*, 1996, p. 93). For instance, concentrating on more recent changes in carbon sinks is likely to favour those regions that have deforested earlier, e.g., Europe.

Cumulative and Historical Emissions

Cumulative anthropogenic (i.e., human-emitted) emissions of CO_2 from fossil fuel use are a major cause of global warming, and give some indication of which countries have contributed most to human-induced climate change.

Top-5 historic CO_2 contributors by region over the years 1800 to 1988 (in %)		
Region	***Industrial CO_2***	***Total CO_2***
OECD North America	33.2	29.7
OECD Europe	26.1	16.6
Former USSR	14.1	12.5
China	5.5	6.0
Eastern Europe	5.5	4.8

The table above to the left is based on Banuri *et al.* (1996, p. 94). Overall, developed countries accounted for 83.8% of industrial CO_2 emissions over this time period, and 67.8% of total CO_2 emissions. Developing countries accounted for industrial CO_2 emissions of 16.2% over this time period, and 32.2% of total CO_2 emissions. The estimate of total CO_2 emissions includes biotic carbon emissions, mainly from deforestation. Banuri *et al.* (1996, p. 94) calculated per capita cumulative emissions based on then-current population. The ratio in per capita emissions between industrialized countries and developing countries was estimated at more than 10 to 1.

Including biotic emissions brings about the same controversy mentioned earlier regarding carbon sinks and land-use change (Banuri *et al.*, 1996, pp. 93–94). The actual calculation of net emissions is very complex, and is affected by how carbon sinks are allocated between regions and the dynamics of the climate system.

Non-OECD countries accounted for 42% of cumulative energy-related CO_2 emissions between 1890–2007. Over this time period, the US accounted for 28% of emissions; the EU, 23%; Russia, 11%; China, 9%; other OECD countries, 5%; Japan, 4%; India, 3%; and the rest of the world, 18%.

Changes Since a Particular Base Year

Between 1970–2004, global growth in annual CO_2 emissions was driven by North America, Asia, and the Middle East. The sharp acceleration in CO_2 emissions since 2000 to more than a 3% increase per year (more than 2 ppm per year) from 1.1% per year during the 1990s is attributable to the lapse of formerly declining trends in carbon intensity of both developing and developed nations. China was responsible for most of global growth in emissions during this period. Localised plummeting emissions associated with the collapse of the Soviet Union have been followed by slow emissions growth in this region due to more efficient energy use, made necessary by the increasing proportion of it that is exported. In comparison, methane has not increased appreciably, and N_2O by 0.25% y^{-1}.

Using different base years for measuring emissions has an effect on estimates of national contributions to global warming. This can be calculated by dividing a country's highest contribution to global warming starting from a particular base year, by that country's minimum contribution to global warming starting from a particular base year. Choosing between different base years of 1750, 1900, 1950, and 1990 has a significant effect for most countries. Within the G8 group of countries, it is most significant for the UK, France and Germany. These countries have a long history of CO_2 emissions.

Annual Emissions

Annual per capita emissions in the industrialized countries are typically as much as ten times the average in developing countries. Due to China's fast economic development, its annual per capita emissions are quickly approaching the levels of those in the Annex I group of the Kyoto Protocol (i.e., the developed countries excluding the USA). Other countries with fast growing emissions are South

Korea, Iran, and Australia. On the other hand, annual per capita emissions of the EU-15 and the USA are gradually decreasing over time. Emissions in Russia and the Ukraine have decreased fastest since 1990 due to economic restructuring in these countries.

Energy statistics for fast growing economies are less accurate than those for the industrialized countries.

For China's annual emissions in 2008, the Netherlands Environmental Assessment Agency estimated an uncertainty range of about 10%.

Top Emitters

Annual: In 2009, the annual top ten emitting countries accounted for about two-thirds of the world's annual energy-related CO_2 emissions.

Top-10 annual energy-related CO_2 emitters for the year 2009		
Country	***% of global total annual emissions***	***Tonnes of GHG per capita***
People's Rep. of China	23.6	5.13
United States	17.9	16.9
India	5.5	1.37
Russian Federation	5.3	10.8
Japan	3.8	8.6
Germany	2.6	9.2
Islamic Rep. of Iran	1.8	7.3
Canada	1.8	15.4
Korea	1.8	10.6
United Kingdom	1.6	7.5

Cumulative

Top-10 cumulative energy-related CO_2 emitters between 1850–2008		
Country	***% of world total***	***Metric tones CO_2 per person***
United States	28.5	1,132.7
China	9.36	85.4
Russian Federation	7.95	677.2
Germany	6.78	998.9
United Kingdom	5.73	1,127.8
Japan	3.88	367
France	2.73	514.9
India	2.52	26.7
Canada	2.17	789.2
Ukraine	2.13	556.4

Embedded Emissions

One way of attributing greenhouse gas (GHG) emissions is to measure the embedded emissions (also referred to as "embodied emissions") of goods that are being consumed. Emissions are usually measured according to production, rather than consumption.

For example, in the main international treaty on climate change (the UNFCCC), countries report on emissions produced within their borders, e.g., the emissions produced from burning fossil fuels.

Under a production-based accounting of emissions, embedded emissions on imported goods are attributed to the exporting, rather than the importing, country.

Under a consumption-based accounting of emissions, embedded emissions on imported goods are attributed to the importing country, rather than the exporting, country.

Davis and Caldeira (2010) found that a substantial proportion of CO_2 emissions are traded internationally. The net effect of trade was to export emissions from China and other emerging markets to consumers in the US, Japan, and Western Europe.

Based on annual emissions data from the year 2004, and on a per-capita consumption basis, the top-5 emitting countries were found to be (in tCO_2 per person, per year): Luxembourg (34.7), the US (22.0), Singapore (20.2), Australia (16.7), and Canada (16.6). Carbon Trust research revealed that approximately 25% of all CO2 emissions from human activities 'flow' (i.e. are imported or exported) from one country to another. Major developed economies were found to be typically net importers of embodied carbon emissions — with UK consumption emissions 34% higher than production emissions, and Germany (29%), Japan (19%) and the USA (13%) also significant net importers of embodied emissions.

Effect of Policy

Governments have taken action to reduce GHG emissions (climate change mitigation). Assessments of policy effectiveness have included work by the Intergovernmental Panel on Climate Change, International Energy Agency, and United Nations Environment Programme. Policies implemented by governments have included national and regional targets to reduce emissions, promoting energy efficiency, and support for renewable energy.

Countries and regions listed in Annex I of the United Nations Framework Convention on Climate Change (UNFCCC) (i.e., the OECD and former planned economies of the Soviet Union) are required to submit periodic assessments to the UNFCCC of actions they are taking to address climate change.

Analysis by the UNFCCC (2011) suggested that policies and measures undertaken by Annex I Parties may have produced emission savings of 1.5 thousand Tg CO_2-eq in the year 2010, with most savings made in the energy sector.

The projected emissions saving of 1.5 thousand Tg CO_2-eq is measured against a hypothetical "baseline" of Annex I emissions, i.e., projected Annex I emissions in the absence of policies and measures. The total projected Annex I saving of 1.5 thousand CO_2-eq does not include emissions savings in seven of the Annex I Parties.

Projections

A wide range of projections of future GHG emissions have been produced. Rogner *et al.* (2007) assessed the scientific literature on GHG projections. Rogner *et al.* (2007) concluded that unless energy policies changed substantially, the world would continue to depend on fossil fuels until 2025–2030.

Projections suggest that more than 80% of the world's energy will come from fossil fuels. This conclusion was based on "much evidence" and "high agreement" in the literature.

Projected annual energy-related CO_2 emissions in 2030 were 40–110% higher than in 2000, with two-thirds of the increase originating in developing countries.

Projected annual per capita emissions in developed country regions remained substantially lower (2.8–5.1 tonnes CO_2) than those in developed country regions (9.6–15.1 tonnes CO_2). Projections consistently showed increase in annual world GHG emissions (the "Kyoto" gases, measured in CO_2-equivalent) of 25–90% by 2030, compared to 2000.

Relative CO_2 Emission from Various Fuels

One litre of gasoline, when used as a fuel, produces 2.32 kg (about 1300 litres or 1.3 cubic meters) of carbon dioxide, a greenhouse gas. One US gallon produces 19.4 lb (1,291.5 gallons or 172.65 cubic feet)

Mass of carbon dioxide emitted per quantity of energy for various fuels		
Fuel name	***CO_2 emitted (lbs/10^6 Btu)***	***CO_2 emitted (g/10^6 J)***
Natural gas	117	50.30
Liquefied petroleum gas	139	59.76
Propane	139	59.76
Aviation gasoline	153	65.78
Automobile gasoline	156	67.07
Kerosene	159	68.36
Fuel oil	161	69.22
Tires/tire derived fuel	189	81.26
Wood and wood waste	195	83.83
Coal (bituminous)	205	88.13
Coal (sub-bituminous)	213	91.57
Coal (lignite)	215	92.43
Petroleum coke	225	96.73
Coal (anthracite)	227	97.59

Life-cycle Greenhouse-gas Emissions of Energy Sources

A literature review of numerous energy sources CO2 emissions by the IPCC in 2011, found that, the CO2 emission value, that fell within the 50th percentile of all total life cycle emissions studies conducted, was as follows.

Lifecycle greenhouse gas emissions by electricity source.[123]		
Technology	***Description***	***50th percentile (g CO2/kWh$_e$)***
Hydroelectric	reservoir	4
Wind	onshore	12
Nuclear	various generation II reactor types	16
Biomass	various	18
Solar thermal	parabolic trough	22
Geothermal	hot dry rock	45
Solar PV	Polycrystaline silicon	46
Natural gas	various combined cycle turbines without scrubbing	469
Coal	various generator types without scrubbing	1001

Removal from the Atmosphere ("Sinks")

Natural Processes: Greenhouse gases can be removed from the atmosphere by various processes, as a consequence of:

- a physical change (condensation and precipitation remove water vapour from the atmosphere).
- a chemical reaction within the atmosphere. For example, methane is oxidized by reaction with naturally occurring hydroxyl radical, OH. and degraded to CO_2 and water vapour

(CO_2 from the oxidation of methane is not included in the methane Global warming potential). Other chemical reactions include solution and solid phase chemistry occurring in atmospheric aerosols.

- a physical exchange between the atmosphere and the other compartments of the planet. An example is the mixing of atmospheric gases into the oceans.
- a chemical change at the interface between the atmosphere and the other compartments of the planet. This is the case for CO_2, which is reduced by photosynthesis of plants, and which, after dissolving in the oceans, reacts to form carbonic acid and bicarbonate and carbonate ions.
- a photochemical change. Halocarbons are dissociated by UV light releasing Cl· and F· as free radicals in the stratosphere with harmful effects on ozone (halocarbons are generally too stable to disappear by chemical reaction in the atmosphere).

Negative Emissions

A number of technologies remove greenhouse gases emissions from the atmosphere. Most widely analysed are those that remove carbon dioxide from the atmosphere, either to geologic formations such as bio-energy with carbon capture and storage and carbon dioxide air capture, or to the soil as in the case with biochar. The IPCC has pointed out that many long-term climate scenario models require large scale manmade negative emissions to avoid serious climate change.

History of Scientific Research

Late 19th century scientists experimentally discovered that N_2 and O_2 do not absorb infrared radiation (called, at that time, "dark radiation") while, on the contrary, water, both as true vapour and condensed in the form of microscopic droplets suspended in clouds, as well as CO_2 and other poly-atomic gaseous molecules, do absorb infrared radiation. It was recognized in the early 20th century that greenhouse gases in the atmosphere made the Earth's overall temperature higher than it would be without them. During the late 20th century, a scientific consensus evolved that increasing concentrations of greenhouse gases in the atmosphere are causing a substantial rise in global temperatures and changes to other parts of the climate system, with consequences for the environment and for human health.

8

Erosion

Erosion is the process by which soil and rock are removed from the Earth's surface by exogenetic processes such as wind or water flow, and then transported and deposited in other locations. While erosion is a natural process, human activities have increased by 10-40 times the rate at which erosion is occurring globally. Excessive erosion causes problems such as desertification, decreases in agricultural productivity due to land degradation, sedimentation of waterways, and ecological collapse due to loss of the nutrient rich upper soil layers. Water and wind erosion are now the two primary causes of land degradation; combined, they are responsible for 84% of degraded acreage, making excessive erosion one of the most significant global environmental problems. Industrial agriculture, deforestation, roads, anthropogenic climate change and urban sprawl are amongst the most significant human activities in regard to their effect on stimulating erosion. However, there are many available alternative land use practices that can curtail or limit erosion, such as terrace-building, no-till agriculture, and revegetation of denuded soils.

Physical Processes

Water Erosion

Rainfall: There are four primary types of erosion that occur as a direct result of rainfall—*splash erosion*,*sheet erosion*, *rill erosion*, and *gully erosion*. Splash erosion is generally seen as the first and least severe stage in the soil erosion process, which is followed by sheet erosion, then rill erosion and finally gully erosion (the most severe of the four).

In splash erosion, the impact of a falling raindrop creates a small crater in the soil, ejecting soil particles. The distance these soil particles travel can be as much as two feet vertically and five feet horizontally on level ground. Once the rate of rainfall is faster than the rate of infiltration into the soil, surface runoff occurs and carries the loosened soil particles down the slope.

Sheet erosion is the transport of loosened soil particles by overland flow.

Rill erosion refers to the development of small, ephemeral concentrated flow paths which function as both sediment source and sediment delivery systems for erosion on hillslopes. Generally, where water erosion rates on disturbed upland areas are greatest, rills are active. Flow depths in rills are typically on the order of a few centimeters or less and slopes may be quite steep. This means that rills exhibit very different hydraulic physics than water flowing through the deeper, wider channels of streams and rivers.

Gully erosion occurs when runoff water accumulates, and then rapidly flows in narrow channels during or immediately after heavy rains or melting snow, removing soil to a considerable depth.

Rivers and Streams

Valley or *stream erosion* occurs with continued water flow along a linear feature. The erosion is both downward, deepening the valley, and headward, extending the valley into the hillside. In the earliest stage of stream erosion, the erosive activity is dominantly vertical, the valleys have a typical V cross-section and the stream gradient is relatively steep. When some base level is reached, the erosive activity switches to lateral erosion, which widens the valley floor and creates a narrow floodplain. The stream gradient becomes nearly flat, and lateral deposition of sediments becomes important as the stream meanders across the valley floor. In all stages of stream erosion, by far the most erosion occurs during times of flood, when more and faster-moving water is available to carry a larger sediment load. In such processes, it is not the water alone that erodes: suspended abrasive particles, pebbles and boulders can also act erosively as they traverse a surface, in a process known as *traction*.

Bank erosion is the wearing away of the banks of a stream or river. This is distinguished from changes on the bed of the watercourse, which is referred to as *scour*. Erosion and changes in the form of river banks may be measured by inserting metal rods into the bank and marking the position of the bank surface along the rods at different times.

Thermal erosion is the result of melting and weakening permafrost due to moving water. It can occur both along rivers and at the coast. Rapid river channel migration observed in the Lena River of Siberia is due to thermal erosion, as these portions of the banks are composed of permafrost-cemented non-cohesive materials. Much of this erosion occurs as the weakened banks fail in large slumps. Thermal erosion also affects the Arctic coast, where wave action and near-shore temperatures combine to undercut permafrost bluffs along the shoreline and cause them to fail. Annual erosion rates along a 100-kilometre segment of the Beaufort Sea shoreline averaged 5.6 metres per year from 1955 to 2002.

Coastal Erosion

Shoreline erosion, which occurs on both exposed and sheltered coasts, primarily occurs through the action of currents and waves but sea level (tidal) change can also play a role. *Hydraulic action* takes place when air in a joint is suddenly compressed by a wave closing the entrance of the joint. This then cracks it. *Wave pounding* is when the sheer energy of the wave hitting the cliff or rock breaks pieces off. *Abrasion* or *corrasion* is caused by waves launching seaload at the cliff. It is the most effective and rapid form of shoreline erosion (not to be confused with *corrosion*). *Corrosion* is the dissolving of rock by carbonic acid in sea water. Limestone cliffs are particularly vulnerable to this kind of erosion. *Attrition* is where particles/seaload carried by the waves are worn down as they hit each other and the cliffs. This then makes the material easier to wash away. The material ends up as shingle and sand. Another significant source of erosion, particularly on carbonate coastlines, is the boring, scraping and grinding of organisms, a process termed *bioerosion.*

Sediment is transported along the coast in the direction of the prevailing current (longshore drift). When the upcurrent amount of sediment is less than the amount being carried away, erosion occurs. When the upcurrent amount of sediment is greater, sand or gravel banks will tend to form as a result of deposition. These banks may slowly migrate along the coast in the direction of the longshore drift, alternately protecting and exposing parts of the coastline. Where there is a bend in the coastline, quite often a build up of eroded material occurs forming a long narrow bank (a spit). Armoured beaches and submerged offshore sandbanks may also protect parts of a coastline from erosion. Over the years, as the shoals gradually shift, the erosion may be redirected to attack different parts of the shore.

Glaciers

Glaciers erode predominantly by three different processes: abrasion/scouring, plucking, and ice thrusting. In an abrasion process, debris in the basal ice scrapes along the bed, polishing and gouging the underlying rocks, similar to sandpaper on wood. Glaciers can also cause pieces of bedrock to crack off in the process of plucking. In ice thrusting, the glacier freezes to its bed, then as it surges forward, it moves large sheets of frozen sediment at the base along with the glacier. This method produced some of the many thousands of lake basins that dot the edge of the Canadian Shield. These processes, combined with erosion and transport by the water network beneath the glacier, leave moraines, drumlins, ground moraine (till), kames, kame deltas, moulins, and glacial erratics in their wake, typically at the terminus or during glacier retreat.

Floods

At extremely high flows, kolks, or vortices are formed by large volumes of rapidly rushing water. Kolks cause extreme local erosion, plucking bedrock and creating pothole-type geographical features called Rock-cut basins. Examples can be seen in the flood regions result from glacial Lake Missoula, which created the channeled scablands in the Columbia Basin region of eastern Washington.

Freezing and Thawing

Cold weather causes water trapped in tiny rock cracks to freeze and expand, breaking the rock into several pieces. This can lead to gravity erosion on steep slopes. The scree which forms at the bottom of a steep mountainside is mostly formed from pieces of rock (soil) broken away by this means. It is a common engineering problem wherever rock cliffs are alongside roads, because morning thaws can drop hazardous rock pieces onto the road.

Wind Erosion

Wind erosion is a major geomorphological force, especially in arid and semi-arid regions. It is also a major source of land degradation, evaporation, desertification, harmful airborne dust, and crop damage—especially after being increased far above natural rates by human activities such as deforestation, urbanization, and agriculture.

Wind erosion is of two primary varieties: *deflation*, where the wind picks up and carries loose soil particles; and *abrasion*, where surfaces are worn down as they are struck by airborne particles carried by wind. Deflation is divided into three categories: (1) *surface creep*, where larger, heavier particles slide or roll along the ground; (2) *saltation*, where particles are lifted a short height into the air, and bounce and saltate across the surface of the soil; and (3) *suspension*, where very small and light particles are lifted into the air by the wind, and are often carried for long distances. Saltation is responsible for the majority (50-70%) of wind erosion, followed by suspension (30-40%), and then surface creep (5-25%).

Wind erosion is much more severe in arid areas, and during times of drought. For example, in the Great Plains, it is estimated that wind erosion soil loss can be as much as 6100 times greater in drought years, than in wet years.

Gravitational Erosion

Mass movement is the downward and outward movement of rock and sediments on a sloped surface, mainly due to the force of gravity. Mass movement is an important part of the erosional process, and is often the first stage in the breakdown and transport of weathered materials in mountainous areas. It moves material from higher elevations to lower elevations where other eroding agents such as streams and glaciers can then pick up the material and move it to even lower elevations. Mass-movement processes are always occurring continuously on all slopes; some mass-movement processes act very slowly; others occur very suddenly, often with disastrous results. Any perceptible down-slope movement of rock or sediment is often referred to in general terms as a landslide. However, landslides can be classified in a much more detailed way that reflects the mechanisms responsible for the movement and the velocity at which the movement occurs. One of the visible topographical manifestations of a very slow form of such activity is a scree slope.

Slumping happens on steep hillsides, occurring along distinct fracture zones, often within materials like clay that, once released, may move quite rapidly downhill. They will often show a spoon-shaped isostatic depression, in which the material has begun to slide downhill. In some cases, the slump is caused by water beneath the slope weakening it. In many cases it is simply the result of poor engineering along highways where it is a regular occurrence.

Surface creep is the slow movement of soil and rock debris by gravity which is usually not perceptible except through extended observation. However, the term can also describe the rolling of dislodged soil particles 0.5 to 1.0 mm in diameter by wind along the soil surface.

Exfoliation

Exfoliation is a type of erosion that occurs when a rock is rapidly heated up by the sun. This results in the expansion of the rock. When the temperature decreases again, the rock contracts, causing pieces of the rock to break off. Exfoliation occurs mainly in deserts due to the high temperatures during the day and cold temperatures at night.

Factors Affecting Erosion Rates

Precipitation and wind speed: Climatic factors include the amount and intensity of precipitation, the average temperature, as well as the typical temperature range, seasonality, wind speed, and storm frequency. In general, given similar vegetation and ecosystems, areas with high-intensity precipitation, more frequent rainfall, more wind, or more storms are expected to have more erosion.

Rainfall intensity is the primary determinant of erosivity, with higher intensity rainfall generally resulting in more erosion. The size and velocity of rain drops is also an important factor. Larger and higher-velocity rain drops have greater kinetic energy, and thus their impact will displace soil particles by larger distances than smaller, slower-moving rain drops.

Soil Structure and Composition

The composition, moisture, and compaction of soil are all major factors in determining the erosivity of rainfall. Sediments containing more clay tend to be more resistant to erosion than those with sand or silt, because the clay helps bind soil particles together. Soil containing high levels of organic materials are often more resistant to erosion, because the organic materials coagulate soil colloids and create a stronger, more stable soil structure. The amount of water present in the soil before the precipitation also plays an important role, because it sets limits on the amount of water that can be absorbed by the soil (and hence prevented from flowing on the surface as erosive runoff). Wet, saturated soils will not be able to absorb as much rain water, leading to higher levels of surface runoff and thus higher erosivity for a given volume of rainfall. Soil compaction also affects the permeability of the soil to water, and hence the amount of water that flows away as runoff. More compacted soils will have a larger amount of surface runoff than less compacted soils.

Vegetative Cover

Vegetation acts as an interface between the atmosphere and the soil. It increases the permeability of the soil to rainwater, thus decreasing runoff. It shelters the soil from winds, which results in decreased wind erosion, as well as advantageous changes in microclimate. The roots of the plants bind the soil together, and interweave with other roots, forming a more solid mass that is less susceptible to both water and wind erosion. The removal of vegetation increases the rate of surface erosion.

Topography

The topography of the land determines the velocity at which surface runoff will flow, which in turn determines the erosivity of the runoff. Longer, steeper slopes (especially those without adequate vegetative cover) are more susceptible to very high rates of erosion during heavy rains than shorter, less steep slopes. Steeper terrain is also more prone to mudslides, landslides, and other forms of gravitational erosion processes.

Human Activities that Increase Erosion Rates

Agricultural Practices: Unsustainable agricultural practices are the single greatest contributor to the global increase in erosion rates. The tillage of agricultural lands, which breaks up soil into finer particles, is one of the primary factors. The problem has been exacerbated in modern times, due to mechanized agricultural equipment that allows for deep plowing, which severely increases the amount of soil that is available for transport by water erosion.

Others include mono-cropping, farming on steep slopes, pesticide and chemical fertilizer usage (which kill organisms that bind soil together), row-cropping, and the use of surface irrigation. A complex overall situation with respect to defining nutrient losses from soils, could arise as a result of the size selective nature of soil erosion events. Loss of total phosphorus, for instance, in the finer eroded fraction is greater relative to the whole soil.

Extrapolating this evidence to predict subsequent behaviour within receiving aquatic systems, the reason is that this more easily transported material may support a lower solution P concentration compared to coarser sized fractions. Tillage also increases wind erosion rates, by dehydrating the soil and breaking it up into smaller particles that can be picked up by the wind. Exacerbating this is the fact that most of the trees are generally removed from agricultural fields, allowing winds to have long, open runs to travel over at higher speeds. Heavy grazing reduces vegetative cover and causes severe soil compaction, both of which increase erosion rates.

Deforestation

In an undisturbed forest, the mineral soil is protected by a layer of *leaf litter* and an *humus* that cover the forest floor. These two layers form a protective mat over the soil that absorbs the impact of rain drops. They are porous and highly permeable to rainfall, and allow rainwater to slow percolate into the soil below, instead of flowing over the surface as runoff. The roots of the trees and plants hold together soil particles, preventing them from being washed away. The vegetative cover acts to reduce the velocity of the raindrops that strike the foliage and stems before hitting the ground, reducing their kinetic energy. However it is the forest floor, more than the canopy, that prevents surface erosion. The terminal velocity of rain drops is reached in about 8 metres. Because forest canopies are usually higher than this, rain drops can often regain terminal velocity even after striking the canopy. However, the intact forest floor, with its layers of leaf litter and organic matter, is still able to absorb the impact of the rainfall.

Deforestation causes increased erosion rates due to exposure of mineral soil by removing the humus and litter layers from the soil surface, removing the vegetative cover that binds soil together, and causing heavy soil compaction from logging equipment. Once trees have been removed by fire or logging, infiltration rates become high and erosion low to the degree the forest floor remains intact. Severe fires can lead to significant further erosion if followed by heavy rainfall.

Globally one of the largest contributors to erosive soil loss in the year 2006 is the slash and burn treatment of tropical forests. In a number of regions of the earth, entire sectors of a country have been rendered unproductive. For example, on the Madagascar high central plateau, comprising approximately ten percent of that country's land area, virtually the entire landscape is sterile of vegetation, with gully erosive furrows typically in excess of 50 metres deep and one kilometre wide. Shifting cultivation is a farming system which sometimes incorporates the slash and burn method in some regions of the world. This degrades the soil and causes the soil to become less and less fertile.

Roads and Urbanization

Urbanization has major effects on erosion processes—first by denuding the land of vegetative cover, altering drainage patterns, and compacting the soil during construction; and next by covering the land in an impermeable layer of asphalt or concrete that increases the amount of surface runoff and increases surface wind speeds. Much of the sediment carried in runoff from urban areas (especially roads) is highly contaminated with fuel, oil, and other chemicals. This increased runoff, in addition to eroding and degrading the land that it flows over, also causes major disruption to surrounding watersheds by altering the volume and rate of water that flows through them, and filling them with chemically polluted sedimentation. The increased flow of water through local waterways also causes a large increase in the rate of bank erosion.

Climate Change

The warmer atmospheric temperatures observed over the past decades are expected to lead to a more vigorous hydrological cycle, including more extreme rainfall events. The rise in sea levels that has occurred as a result of climate change has also greatly increased coastal erosion rates.

Studies on soil erosion suggest that increased rainfall amounts and intensities will lead to greater rates of erosion. Thus, if rainfall amounts and intensities increase in many parts of the world as expected, erosion will also increase, unless amelioration measures are taken. Soil erosion rates are expected to change in response to changes in climate for a variety of reasons.

The most direct is the change in the erosive power of rainfall. Other reasons include: a) changes in plant canopy caused by shifts in plant biomass production associated with moisture regime; b) changes in litter cover on the ground caused by changes in both plant residue decomposition rates driven by temperature and moisture dependent soil microbial activity as well as plant biomass production rates; c) changes in soil moisture due to shifting precipitation regimes and evapo-transpiration rates, which changes infiltration and runoff ratios; d) soil erodibility changes due to decrease in soil organic matter concentrations in soils that lead to a soil structure that is more susceptible to erosion and increased runoff due to increased soil surface sealing and crusting; e) a shift of winter precipitation from non-erosive snow to erosive rainfall due to increasing winter temperatures; f) melting of permafrost, which induces an erodible soil state from a previously non-erodible one; and g) shifts in land use made necessary to accommodate new climatic regimes.

Studies by Pruski and Nearing indicated that, other factors such as land use not considered, it is reasonable to expect approximately a 1.7% change in soil erosion for each 1% change in total precipitation under climate change.

Global Environmental Effects

Due to the severity of its ecological effects, and the scale on which it is occurring, erosion constitutes one of the most significant global environmental problems we face today.

Land Degradation

Water and wind erosion are now the two primary causes of land degradation; combined, they are responsible for 84% of degraded acreage. Each year, about 75 billion tons of soil is eroded from the land—a rate that is about 13-40 times as fast as the natural rate of erosion. Approximately 40% of the world's agricultural land is seriously degraded. According to the United Nations, an area of fertile soil the size of Ukraine is lost every year because of drought, deforestation and climate change. In Africa, if current trends of soil degradation continue, the continent might be able to feed just 25% of its population by 2025, according to UNU's Ghana-based Institute for Natural Resources in Africa. The loss of soil fertility due to erosion is further problematic because the response is often to apply chemical fertilizers, which leads to further water and soil pollution, rather than to allow the land to regenerate.

Sedimentation of Aquatic Ecosystems

Soil erosion (especially from agricultural activity) is considered to be the leading global cause of diffuse water pollution, due to the effects of the excess sediments flowing into the world's waterways. The sediments themselves act as pollutants, as well as being carriers for other pollutants, such as attached pesticide molecules or heavy metals. The effect of increased sediments loads on aquatic ecosystems can be catastrophic. Silt can smother the spawning beds of fish, by filling in the space between gravel on the stream bed. It also reduces their food supply, and causes major respiratory issues for them as sediment enters their gills. The biodiversity of aquatic plant and algal life is reduced, and invertebrates are also unable to survive and reproduce. While the sedimentation event itself might be relatively short-lived, the ecological disruption caused by the mass die off often persists long into the future.

One of the most serious and long-running water erosion problems worldwide is in the People's Republic of China, on the middle reaches of the Yellow River and the upper reaches of the Yangtze River. From the Yellow River, over 1.6 billion tons of sediment flows into the ocean each year. The sediment originates primarily from water erosion in the Loess Plateau region of the northwest.

Monitoring, Measuring, and Modelling Erosion

Monitoring and modelling of erosion processes can help us better understand the causes, make predictions, and plan how to implement preventative and restorative strategies. However, the complexity of erosion processes and the number of areas that must be studied to understand and model them (e.g. climatology, hydrology, geology, chemistry, physics, etc.) makes accurate modelling quite challenging. Erosion models are also non-linear, which makes them difficult to work with numerically, and makes it difficult or impossible to scale up to making predictions about large areas from data collected by sampling smaller plots.

Thc most commonly uscd modcl for prcdicting soil loss from water erosion is the *Universal Soil Loss Equation (USLE)*, which estimates the average annual soil loss Aas:

$$A = RKLSCP$$

where R is the rainfall erosivity factor, $K$is the soil erodibility factor, L and S are topographic factors representing length and slope, and C and P are cropping management factors.

Erosion is measured and further understood using tools such as the micro-erosion metre (MEM) and the traversing micro-erosion metre (TMEM). The MEM has proved helpful in measuring bedrock erosion in various ecosystems around the world. It can measure both terrestrial and oceanic erosion. On the other hand, the TMEM can be used to track the expanding and contracting of volatile rock formations and can give a reading of how quickly a rock formation is deteriorating.

Prevention and Remediation

The most effective known method for erosion prevention is to increase vegetative cover on the land, which helps prevent both wind and water erosion. Terracing is an extremely effective means of erosion control, which has been practiced for thousands of years by people all over the world. Windbreaks (also called shelterbelts) are rows of trees and shrubs that are planted along the edges of agricultural fields, to shield the fields against winds. In addition to significantly reducing wind erosion, windbreaks provide many other benefits such as improved microclimates for crops (which are sheltered from the dehydrating and otherwise damaging effects of wind), habitat for beneficial bird species, carbon sequestration, and aesthetic improvements to the agricultural landscape. Traditional planting methods, such as mixed-cropping (instead of monocropping) and crop rotation have also been shown to significantly reduce erosion rates.

Erosion Control

Erosion control is the practice of preventing or controlling wind or water erosion in agriculture, land development and construction. Effective erosion controls are important techniques in preventing water pollution and soil loss.

Erosion controls are used in natural areas, agricultural settings or urban environments. In urban areas erosion controls are often part of stormwater runoff management programmes required by local governments. The controls often involve the creation of a physical barrier, such as vegetation or rock, to absorb some of the energy of the wind or water that is causing the erosion. On construction sites they are often implemented in conjunction with sediment controls such as sediment basins and silt fences. Bank erosion is a natural process: without it, rivers would not meander and change course. However, land management patterns that change the hydrograph and/or vegetation cover can act to increase or decrease channel migration rates. In many places, whether or not the banks are unstable due to human activities, people try to keep a river in a single place. This can be done for environmental reclamation or to prevent a river from changing course into land that is being used by people. One way that this is done is by placing riprap or gabions along the bank (Bank Erosion by Wikipedia).

Mathematical Modelling

Since the 1920s and 1930s scientists have been creating mathematical models for understanding the mechanisms of soil erosion and resulting sediment surface runoff, including an early paper by Albert Einstein applying Baer's law. These models have addressed both gully and sheet erosion. Earliest models were a simple set of linked equations which could be employed by manual calculation. By the 1970s the models had expanded to complex computer models addressing nonpoint source pollution with thousands of lines of computer code. The more complex models were able to address nuances in micrometreology, soil particle size distributions and micro-terrain variation.

Bibliography

Ashman, K.: *Seeds of Change: Strategies for Food Security for the Inner City.* University of California, 1993.

Ben Wisner: *At Risk, Natural Hazards, People's Vulnerability, and Disasters*, London, Routledge, 1994.

Brij K. Taimni: *Food Security in 21 Century : Perspective and Vision*, Konark, Delhi, 2001.

Chakraborty, D, Das: *Proceedings of Workshop on Water Resources Development of West Bengal*, Jadavpur University, Kalcutta, 1993.

Cristobal Noe Aguilar *: Food Science and Food Biotechnology in Developing Countries*, Asiatech Pub, 2008.

Darst, Robert G. : *Smokestack Diplomacy: Cooperation and Conflict in East- West Environmental Politics,* Cambridge, MA: MIT, 2001.

Das, D. : *Geography and Environmental Pollution*, Concept, New Delhi, 1997.

Debashis Basu; B Francis Kulirani and B Datta Ray: *Agriculture Food Security Nutrition and Health in North East India*, Mittal Pub, Delhi, 2006.

Epstein, S. J. M.: *The Earthy Soil: Bombay Peasants and the Indian Nationalist Movement, 1919-1947*, Oxford University Press, Delhi, 1988.

Ghose, Arpita : *Globalization, Agricultural Growth and Food Security in India*, Deep and Deep, Delhi, 2011.

Ghost, K.C.: *Famines in Bengal, 1170-1943,* Indian Associated Publishing, Calcutta, 1944.

Hanumantha Rao, C.H.: *Agriculture, Food Security, Poverty and Environment*, Oxford University Press, Delhi, 2006.

Hays, J.N.: *Epidemics and Pandemics: Their Impacts on Human History*, Santa Barbara, Calif.: ABC-CLIO, 2005.

Jackson, Peter: *Maps of Meaning: An Introduction to Cultural Geography.* London and Boston: Unwin Hyman, 1989.

Mukhopadhyay, S.P. and A, & Biswas, A.B. : *Studies on some Problems of Atmospheric Pollution in Southern Part of West Bengal*, Centre for Study of Man and Environment, Calcutta, 1985.

Oliver, John E.: *Encyclopedia of World Climatology*, Springer, Delhi, 2005.

Pavaskar Madhoo and Ghosh Nilanjan: *Debate Over Climate Change and Global Warming*, TAER, Delhi, 2011.

Rao, C.H. Hanumantha : *Agriculture, Food Security, Poverty and Environment : Essays on Post-Reform India*, Oxford University Press, Delhi, 2005.

Sujata K. Dass: *Biotechnology and Food Security*, Isha Books, Delhi, 2004.

Thomas Bernauer *: Genes, Trade and Regulation : The Seeds of Conflict in Food Biotechnology*, Universities Press, 2005.

Tribedi, S. : *Proceedings of National Convention of Environment in India, Challenge for the 21st Century*, India, 1994.

Varley, Anne: *Disaster, Development Environments*, New York, J. Wiley, 1994.

Wahrhaftig, Clyde: *A Streetcar to Subduction: and Other Plate Tectonic Trips by Public Transportation in San Francisco*, American Geophysical Union, Washington, D.C., 1984.

Watkins, Eric: *The Middle Eastern Environment*, London, The British Society for Middle Eastern Studie, 1995.

Index

H

I

L

M

N

O

P

R

S

T

U

V

W

❑❑❑